365初中英语天天阅读

365 INVENTIONS THAT CHANGED THE WORLD I

It's all about humans behind each historical invention.

365个改变世界的发明 上册

每一个发明背后，都是人的故事！

语言难度 ★★★★☆
脑力指数 ★★★★☆
趣味指数 ★★★★★

上海教育出版社
SHANGHAI EDUCATIONAL PUBLISHING HOUSE

Copyright Om Books International, India
Copyright Artworks Om Books International
Originally Published in English by Om Books International
This edition revised and published by SHANGHAI EDUCATIONAL PUBLISHING HOUSE CO., LTD Copyright ©2020

本书由上海教育出版社取得授权并改编,未经许可,不得以任何方式复制或抄袭本书部分或全部内容。版权所有,侵权必究。

图书在版编目(CIP)数据

365个改变世界的发明.上册 /《365初中英语天天阅读》编写组编写. — 上海:上海教育出版社,2020.7
(365初中英语天天阅读)
ISBN 978-7-5720-0059-1

Ⅰ.①3… Ⅱ.①3… Ⅲ.①英语—阅读教学—初中—课外读物 Ⅳ.①G634.413

中国版本图书馆CIP数据核字(2020)第103245号

策划编辑　倪雅菁　王雪婷
责任编辑　谈潇潇　倪雅菁
封面设计　朱博韡

365初中英语天天阅读·365个改变世界的发明(上册)
本书编写组　编写

出版发行	上海教育出版社有限公司
官　　网	www.seph.com.cn
地　　址	上海永福路123号
邮　　编	200031
印　　刷	浙江广育爱多印务有限公司
开　　本	889×1194　1/16　印张12.5
字　　数	323千字
版　　次	2020年7月第1版
印　　次	2020年7月第1次印刷
书　　号	ISBN 978-7-5720-0059-1/G·0047
定　　价	88.00元

如发现质量问题,请向出版社调换。电话:021-64377165

前言

提到365，同学们多半会感到十分亲切或者心里泛起阵阵温暖的涟漪——那是童年365个温馨的夜晚，长辈的轻声细语把我们带入多彩的童话世界，让我们领略世界的缤纷和奇幻，激起我们对未来的无限憧憬。

那么说起"365初中英语天天阅读"，又是什么呢？它们是：《365个世界各地好故事》《365个改变世界的发明》《365个惊奇问与答》《365个人体奥秘》和《365个科技真相》。

"365初中英语天天阅读"系列，讲述各国的故事和寓言，展示多元的世界文化；梳理从远古时期到现今改变世界和人们生活的各种发明，展示人类社会如何发展；挖掘历史、文化、生物、医药等领域的有趣知识和意想不到的问题，阐明科学原理；深入人体的器官和细胞，揭示我们身体的奥秘；探索科学与技术领域的各种难题与谜题，展现令人惊叹的科技真相。

"An apple a day keeps the doctor away.（每日一苹果，医生离你远。）"。"365初中英语天天阅读"系列就像"每日一苹果"，是特别为广大初中学生精心准备的课外英语趣味读物。每天读一篇，坚持一整年，不仅能提升你的英语词汇量和阅读水平，而且能开阔你的视野，让你成为百科知识或故事达人。

"365初中英语天天阅读"系列包含五套10册。每套分为上、下两册，共由365篇英语短文构成。每篇短文，配有两个小问题，用以检测对短文的理解并启发同学们发表自己的见解。书后还配有"阅读小帮手"——"参考答案""生词与短语""索引"和"专有名词（与术语）"等附录。同学们可以先阅读短文，借助生词表加深理解，再思考问题，大胆说出自己的想法。

本套《365个改变世界的发明》以年代为线索，不拘泥于重大发明，用365篇短文介绍了365个与我们的日常生活和各行各业息息相关的事物。每篇短文约100—200词，通过追溯器物发展史和相关发明者的趣闻轶事，带领青少年读者了解人类社会的发展进程，拥抱世界文明。**上册**从260万年前谈至19世纪中叶，纵贯各个历史时期，探寻世界各地史前遗迹的古朴、四大文明古国的光辉、文艺复兴的觉醒、地理大发现时代的冒险进取和两次工业革命的变革与突破；**下册**继续谈至我们生活的时代，可窥掘金潮的狂热、两次世界大战危机中的生命力、第三次工业革命的颠覆与赋能……

阅读《365个改变世界的发明（上册）》之前，你能回答下面的问题吗？

◇ 除了使用香料和醋，古埃及人还发现了什么方法来防腐？

◇ 现代人用来美妆的眼线，在古代还有其他功能吗？

◇ 邮政服务为什么在中国汉朝迅速发展起来？

◇ 美洲原住民起初能接受欧洲人的油炸烹饪吗？

◇ 螺丝刀最早就是用来起螺丝的吗？

◇ 为何联合收割机冠名"联合"？

……

同学们是否已经按捺不住自己的好奇心，迫切想知道答案了？那就请翻开《365个改变世界的发明（上册）》，快乐阅读吧！

一年365天，每天读一篇，"用功不求太猛，但求有恒"。

《老子》有云，"合抱之木，生于毫末；九层之台，始于累土；千里之行，始于足下"。法国谚语也说，Petit à petit, l'oiseau fait son nid.（一点又一点，小鸟筑成巢。）

相信365天之后，你一定会感受到量变到质变的飞跃，英语语言水平的飞跃，知识和见解的飞跃，坚持不懈的毅力和自我管理效能的飞跃。你一定会收获一个全新的你，一个不同凡响的自己！你的努力和进步，大家都会看见！

本书编写组

Contents（目录）

2.6 million years ago – 3000 BC
（260万年前–公元前3000年）

1. Stone Tool（石器） 1
2. Fire（火） 2
3. Knife（刀） 3
4. Log Raft（木筏） 3
5. Spear（矛） 4
6. Fur（皮草） 4
7. Brick（砖块） 5
8. Bow and Arrow（弓箭） 5
9. Cave Painting（岩画） 6
10. Flute（笛子） 7
11. Sewing Needle（缝衣针） 7
12. Statue（雕像） 8
13. Rope（绳子） 8
14. Pigment（颜料） 9
15. Preservative（防腐剂） 9
16. Fishing Net（渔网） 10
17. Pottery（陶艺） 10
18. Language（语言） 11
19. Fermentation（发酵） 12

20. Pillow（枕头） 12
21. Mortar（砂浆） 13
22. Refined Salt（精制盐） 13
23. Axe（斧） 14
24. Irrigation（灌溉） 14
25. Leather（皮革） 15
26. Glue（胶水） 16
27. Papyrus（纸莎草） 16
28. Shoes（鞋） 17
29. Sundial（日晷） 17
30. Wheel（轮子） 18
31. Belt（皮带） 18
32. Ship（船） 19
33. Saw（锯子） 20
34. Cement（水泥） 20
35. Bronze（青铜） 21
36. Button（纽扣） 21
37. Silk（丝绸） 22
38. Wig（假发） 23
39. Pen（钢笔） 23
40. Kohl（眼线） 24
41. Shadoof（桔槔） 24
42. Iron（铁） 25
43. Toothpaste（牙膏） 26
44. Weighing Scale（天平） 27
45. Ploughshare（犁铧） 27
46. Dentistry（牙科） 28
47. Pants（裤子） 29
48. Paved Road（铺面道路） 29

2999 BC – 1 BC
（公元前2999年–公元前1年）

49. Sewage System（下水系统） 30
50. Concrete（混凝土） 31
51. Plier（钳子） 32
52. Dam（水坝） 32
53. Soap（肥皂） 33
54. Toilet（厕所） 33
55. Dye（染料） 34
56. Column（柱子） 34
57. Baking（烘焙） 35

85.	Candle（蜡烛）	51
86.	Lever（杠杆）	52
87.	Dome（圆屋顶）	52
88.	Milling（磨盘）	53
89.	Traditional Chinese Calendar（传统中国农历）	53
90.	Oven（炉子）	54
91.	Postal Service（邮政服务）	55
92.	Swimsuit（泳衣）	55
93.	Socks（袜子）	56

1 AD – 1600 AD
（公元1年-公元1600年）

58.	Frying（油炸）	35
59.	Plough（犁）	36
60.	Perfume（香水）	36
61.	Architectural Arch（拱门）	37
62.	Ruler（尺子）	38
63.	Umbrella（伞）	38
64.	Chariot（马车）	39
65.	Scissors（剪刀）	39
66.	Refined Sugar（精制糖）	40
67.	Steel（钢）	40
68.	Archimedes' Screw（阿基米德螺旋泵）	41
69.	Scarf（围巾）	42
70.	Shower（淋浴）	42
71.	Gloves（手套）	43
72.	Bathtub（浴缸）	43
73.	Lock and Key（锁具）	44
74.	Calliper（测径规）	44
75.	Aqueduct（高架渠）	45
76.	Catapult（石弩）	46
77.	Barrel（桶）	46
78.	Magnifying Glass（放大镜）	47
79.	Wheelbarrow（独轮车）	47
80.	Pulley（滑轮）	48
81.	Waterwheel（水车）	49
82.	Plumbing（管道系统）	50
83.	Compass（指南针）	50
84.	Palanquin（轿子）	51

94.	Paper（纸）	57
95.	Abacus（算盘）	58
96.	Toothbrush（牙刷）	58
97.	Vault（拱顶）	59
98.	Hydrometer（液体比重计）	59
99.	Map（地图）	60
100.	Paper Money（纸币）	61
101.	Cannon（大炮）	61
102.	Pretzel（椒盐卷饼）	62
103.	Gun（枪）	63
104.	Gunpowder（火药）	63
105.	Velvet（法兰绒）	64
106.	Rocket（火箭）	64
107.	Dress（裙子）	65
108.	Spinning Wheel（纺车）	66
109.	Rifle（来福枪）	67
110.	Lace（蕾丝）	67
111.	Spectacles（眼镜）	68

112. Screwdriver（螺丝刀）		69
113. Electricity（电）		69
114. Pencil（铅笔）		70
115. Watch（表）		70
116. Printing Press（印刷术）		71
117. Corset（紧身衣）		72
118. Teapot（茶壶）		72

119. Microscope（显微镜）		73
120. Heeled Shoes（高跟鞋）		74
121. Stockings（长筒袜）		74
122. Bullet（子弹）		75
123. Hat（帽子）		75
124. Clothes Iron（熨铁）		76
125. Gregorian Calendar（格里高利历）		76
126. Thermometer（温度计）		77

1601 AD – 1800 AD （公元1601年–公元1800年）

127. Railroad（铁路）		78
128. Cork（软木塞）		79
129. Bow Tie（领结）		79
130. Submarine（潜水艇）		80
131. Steam Engine（蒸汽机）		80
132. Telescope（望远镜）		81
133. Tie（领带）		82
134. Barometer（气压计）		82
135. Blood Transfusion（输血）		83
136. Parachute（降落伞）		84
137. Refrigerator（冰箱）		85
138. Water Frame（水力纺纱机）		85
139. Razor（剃须刀）		86
140. Mayonnaise（蛋黄酱）		86
141. Accelerometer（加速度计）		87
142. Carbonated Water（气泡水）		87

143.	Spinning Jenny（珍妮纺纱机）	88
144.	Sandwich（三明治）	89
145.	Smallpox Vaccine（天花疫苗）	90
146.	Steam Boat（蒸汽船）	90
147.	Iron Bridge（铁桥）	91
148.	Hot Air Balloon（热气球）	92
149.	Thresher（脱粒机）	93
150.	Cotton Gin（轧棉机）	93
151.	Ball Bearing（滚珠轴承）	94

152.	Primary Cell Battery（原电池）	95

1801 AD – 1850 AD
（公元1801年–公元1850年）

153.	Protractor（量角器）	96
154.	Hang Glider（滑翔机）	97
155.	Quinine（奎宁）	97
156.	Tin Can（锡罐）	98
157.	Solar Cell（太阳能电池）	98
158.	Tractor（拖拉机）	99
159.	Reaper（收割机）	100
160.	Stethoscope（听诊器）	100
161.	Camera（照相机）	101

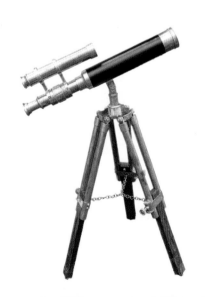

162.	Bicycle（自行车）	102
163.	Suspenders（吊裤带）	103
164.	Matchstick（火柴）	103
165.	Macintosh Raincoat（麦金托什雨衣）	104
166.	Bus（公交车）	104
167.	Braille（布拉耶盲文）	105
168.	Sewing Machine（缝纫机）	106
169.	Handbag（手提包）	107
170.	Harvester（联合收割机）	107
171.	Morse Code（摩斯密码）	108
172.	Telegraph（电报）	109
173.	Stamps（邮票）	109

174.	Suspension Bridge（悬索桥）	110
175.	Stapler（订书机）	111
176.	Aluminium（铝）	111
177.	Voltmeter（电压表）	112
178.	Dirigible（飞艇）	113
179.	Safety Pin（别针）	113
180.	Fax（传真）	114
181.	Elevator（厢式电梯）	114
182.	General Anaesthesia（全身麻醉）	115
183.	Antiseptics（抗菌剂）	115

Possible Answers（参考答案）	116
Words and Expressions（生词与短语）	129
Index（索引）	172
Proper Nouns and Terms（专有名词与术语）	185

2.6 million years ago – 3000 BC
（260万年前–公元前3000年）

1. Stone Tool（石器）

According to Charles Darwin's theory of natural selection, human beings evolved from an ape-like ancestor around six million years ago. As their intellectual ability and physical appearance started changing, humans also began to create tools by using materials from their surroundings. This ability to create and use tools is what differentiated humans from other animals.

The earliest forms of toolmaking are said to have evolved around 2.6 million years ago. All these tools were made of stone, which is why this entire period is called the Stone Age. The Stone Age is divided into three different eras, depending on the kind of tools that were used in each era. These eras are the Palaeolithic Age or the Old Stone Age, the Mesolithic Age or the Middle Stone Age, and the Neolithic Age or the New Stone Age.

The oldest stone tools are known as the "Oldowan[①]" toolkit. They include hammer stones, stone cores and sharp stone flakes.

1. According to the passage, what is the "Oldowan" toolkit?
2. What's the importance of stone tools?

---
① Oldowan 奥尔杜韦文化的

2 Fire（火）

Controlled fire is one of the earliest human discoveries and a crucial one that aided the evolution of humans. Once humans could control fire, they could use it to generate light and heat, to clear forests for farming, to create ceramic objects out of clay and to aid the making of stone tools.

The earliest evidence of controlled fire dates back to the Early Stone Age. Archaeologists unearthed evidence in the form of charred wood and seeds at the Gesher Benot Ya'aqov site① in Israel. These are estimated to be about 790,000 years old!

Some scientists dispute this by stating that the wood and seeds were not an evidence of controlled fire, but of natural fires that humans took advantage of. However, there may be indirect evidence that supports the discovery of controlled fire 700,000 years ago. Around that time, the human brain started growing larger and developing in a way that would be impossible without cooked food. Therefore, scientists conclude that humans must have discovered a way to control fire by then.

> 1. According to the passage, what was the use of fire in the past?
> 2. Fire can help us in many ways, but it can be dangerous as well. What may happen if you are not careful with fire?

---

① Gesher Benot Ya'aqov site 格瑟尔·贝诺·雅科夫遗址

2.6 million years ago – 3000 BC
（260万年前-公元前3000年）

3 Knife（刀）

Knives were first made by banging rocks repeatedly together in such a way that one of the rocks eventually obtained the desired shape. These knives were very crude and rudimentary. As they were made out of stone, these ancient knives were far from the sharp knives that we use today.

During the Middle Ages, as technology improved and the humans began using steel, more impressive blades were made. They were used as swords during the 13th and 14th centuries. Slowly, these swords were refined and made smaller, eventually making their way to our dining tables!

> 1. According to the passage, what were ancient knives made out of?
> 2. How were ancient knives changed into the ones we use on our dining tables?

4 Log Raft（木筏）

The first boat to travel through the sea was discovered around 800,000 years ago. Back then, humans had begun to evolve, but they could not communicate. Scientists called them Homo erectus[①]. Homo erectus evolved in Africa and slowly started migrating to other parts of the Earth. These humans travelled to Indonesia, crossing many leagues at sea. This is when they discovered the log raft.

Knowing what plants and tools were available at the time, scientists have tried to guess and reconstruct the watercraft they might have used. Scientists believed that they used giant bamboos that grew in the region.

> 1. According to the passage, when did Homo erectus discover the log raft?
> 2. Since boats and ships are widely used nowadays, do you think log rafts will disappear in the future? Give your reasons.

---

① Homo erectus 直立人

5 Spear（矛）

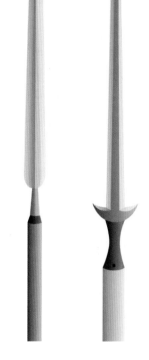

The development of the spear is considered to be the greatest technological feat in the history of humankind. Engineered almost 500,000 years ago in South Africa's Northern Cape province, the spear was developed by the Homo heidelbergensis① species of humans. This species used a wooden shaft as the handle of the spear; a hand-chiselled stone as the weapon and mixed adhesives to hold the spear together.

Interestingly, scientists consider the development of the spear as a testimony to the evolving complex reasoning and power of humankind. However, recent studies show that this invention marked an era of peace back then. Before the discovery, early humans acquired territories using violence. When one group invented spears, the other group grew cautious and avoided fighting, which resulted in less tension among communities.

1. According to the passage, how did the Homo heidelbergensis species of humans develop the spear?
2. Do you think the spear is an important invention? Why?

6 Fur（皮草）

Have you ever imagined what the primitive humans wore? Did they wear trendy clothes like us? What did they wear when they hunted in the early days? The answer to all these questions is fur. Yes, you read that right!

Fur, which has become a controversial fashion symbol today, was once the only thing that humans wore. Around 70,000 years ago, Neanderthals② lived in places with volatile climates. Temperatures changed drastically from very hot to very cold and they had to find a way to protect themselves against the Arctic cold. This led to the discovery of animal skin and its use as clothing. Skins of hairy mammoths, bears, deer and musk oxen were used to make clothes.

1. According to the passage, why was fur invented in the past?
2. Why has fur become a controversial fashion symbol today?

---

① Homo heidelbergensis 海德堡人　② Neanderthals 尼安德特人

2.6 million years ago – 3000 BC
（260万年前—公元前3000年）

7 Brick（砖块）

Just as civilisation was shaping up, humans felt the need to build houses. We know that most houses are made of bricks. Were these bricks also used by primitive cultures like the Nile, the Tigris and Euphrates, the Indus, and the Yellow River? Not exactly! Around 8500 BC, during the Bronze Age, humans started making houses of mud bricks.

Mud bricks were made using the clay found on riverbanks. This clay was mixed with straw, kept in wooden moulds and left to dry. After they dried completely, the bricks were left in the sun to bake naturally. Mesopotamian civilisation saw the first kiln-fired bricks. Mesopotamian masonry technology helped build great structures like the temple at Tepe Gawra① and the ziggurats at Ur and Borsippa (Birs Nimrud). These buildings were up to 87 feet high.

1. When did humans start making houses of mud bricks?
2. What's the difference between mud bricks and kiln-fired bricks?

8 Bow and Arrow（弓箭）

Scientific evidence suggests that the formidable bow and arrow existed around 64,000 years ago in Sibudu Cave②, South Africa. However, it is believed that this intelligent invention is much older. The bow and arrow served as both a military weapon and a hunting tool.

1. According to the passage, what did the bow and arrow serve as?
2. Nowadays we have weapons which are stronger than the bow and arrow. Do you think the bow and arrow will disappear in the future?

Over the years, different cultures used the bow and arrow for different purposes. For instance, in the Mediterranean region, Europe, China, Japan and Eurasian countries it was used as a military tool. While Europe saw the invention of the crossbow and the English came up with the longbow, the Huns, Seljuk Turks③, and Mongols were brilliant, mounted archers. They used recurved bows, which were made of wood and the horns of animals. On the other hand, North American Indians, Inuit and Africans used the regular bow and the crossbow, primarily for hunting.

① Tepe Gawra 高拉土丘　② Sibudu Cave 诗巫渡洞　③ Seljuk Turks 塞尔柱土耳其人

9 Cave Painting（岩画）

Cave paintings can be considered as the primary evidence of the evolution of the human brain. The first cave painting dates back to around 40,000 years in the last phase of the Stone Age. It is also called the Upper Palaeolithic era. Researchers believe that cave art originated in the Aurignacian period in Germany and reached its apex in the late Magdalenian period in France.

Also known as rock paintings, cave paintings were believed to be made by the elders or shamans of the tribe. Early cave paintings were representations of wild animals like deer, bison, horses and aurochs. Abstract patterns and tracings of human hands were also common. It is believed that shamans would retire to the insides of caves to think and draw images of anything that occurred to them.

As opposed to the vast colour palettes and options available for sketching, drawing and painting today, only three colours were available during prehistoric times — red, black and yellow. Even the drawing tools were small and sharpened stones used to scrape on the surface of caves. Iron oxide① was used as red paint, manganese oxide② as black, and clay and yellow ochre③ were used as yellow paint.

1. According to the passage, what did the early cave paintings represent?
2. Some people think cave paintings are treasures from ancient times. Do you think so?

---

① iron oxide 氧化铁 ② manganese oxide 氧化锰 ③ ochre 赭石

10 Flute（笛子）

Most people love musical instruments. Some like pianos; some like guitars and others like drums. Did you ever wonder how old these instruments are? One such instrument — the flute — is around 43,000 years old!

The oldest existing flute was discovered in 2008 in the Hohle Fels Cave[①] near Ulm, Germany. It was made of the bone of a Griffon Vulture[②]. It measured around 8.5 inches in length. Evidence of ancient flutes has been found in ancient Greece, Etruria, India, China and Japan. The flutes that were found later were made of boxwood and had six finger holes. Today, flutes are made of metal as well. Also called the Western Concert, these are generally made of brass and covered in silver.

1. According to the passage, what was the oldest existing flute made of?
2. Do you think people in ancient China played the flute? Give your reasons.

11 Sewing Needle（缝衣针）

The first evidence of the sewing needle was found in South Africa. According to researchers, the needle dates back to around 30,000 years.

The first version of the sewing needle was made of bone. Instead of having an eye to sew the thread, this needle had a closed hook. Later versions of the needle were made of wood, ivory and eventually, steel and plastic. Interestingly, the earlier versions were hook-shaped instead of the straight ones that we see today.

1. According to the passage, what was the first version of the sewing needle like?
2. Why did people change the shape of the sewing needle? Give your reasons.

---

① Hohle Fels Cave 霍赫勒菲尔斯洞　② Griffon Vulture 秃头鹫

12 Statue（雕像）

You may have seen statues in your neighbourhood, but do you know when the first statue was made?

According to researchers and historians, the oldest statue is 40,000 years old. It was found in the Swabian Alps of Germany and was called "The Lion Man". Another statue, called the "Venus of Hohle Fels", was also found later in the same region. Statues are known to possess great cultural significance. Most statues of China are of Buddhas. One such example is the Leshan Giant Buddha①. Also, the Seven Wonders of the ancient world include the Colossus of Rhodes② and the Statue of Zeus③ at Olympia.

1. According to the passage, what statue was found in the Swabian Alps of Germany?
2. Why did the Chinese build the statue of the Leshan Giant Buddha?

13 Rope（绳子）

The invention of rope laid the foundation of more complex connecting systems like cables. Can you guess how old the rope is?

The first ever rope is believed to be made 28,000 years ago. Interestingly, Egyptians used water reeds, grass leather and animal hair to make ropes for the construction of the Great Pyramids, which continue to stand strong even today.

Back then, people twisted or braided ropes using simple hand tools like sticks and rocks. "The Spinner" was an ancient method used by rope makers. It involved tying a rock at the end of a stick and swinging it around to weave the rope.

1. According to the passage, what does "the Spinner" refer to?
2. Were the ropes made of water reeds, grass leather and animal hair strong? Give your reasons.

① Leshan Giant Buddha 乐山大佛　② Colossus of Rhodes 罗得斯岛上的太阳神巨像　③ Statue of Zeus 宙斯神像

14 Pigment（颜料）

The use of colour or pigment began with cave paintings around 40,000 BC. While the use of colour in cave paintings was minimal, the proper manufacturing and usage of pigments began in the Egyptian civilisation 15,000 years ago. Natural colours were washed, cleaned, purified and crushed to enhance their longevity and pigmentation. Egyptian Blue was one of the most famous colours during that time. It was later replaced by smalt and then cobalt.

The Chinese developed vermillion almost at the same time as the Egyptians developed Egyptian Blue. Two thousand years later, the Romans started using vermillion. The major contribution of the Greeks to the palette was the manufacturing of white lead, which was used as white paint. It is said to be one of the finest pigments ever made.

1. According to the passage, why were natural colours washed, cleaned, purified and crushed?
2. Why is the white lead said to be one of the finest pigments ever made?

15 Preservative（防腐剂）

Most food perishes quickly and loses its taste. To avoid this, preservatives are used. The earliest method of preservation was holding meat above smoke that came from a fire. The smoke would help to dry out the meat and ward off bacteria. Drying fruits and vegetables in the sun was also commonly practised.

Ancient Egyptians used spices and vinegar. They discovered the art of preservation in a strange way. When they buried their dead in the hot desert sand, they noticed that sand would dehydrate the bodies and preserve the flesh. They began using this method with food and called it "drying".

The use of sugar and salt in large quantities also keeps unwanted microorganisms at bay. Jams are an example of this, as are salted meat and fish.

1. According to the passage, what was the earliest method of preservation?
2. If a dead camel was buried in the hot desert sand, would its body decompose quickly? Why?

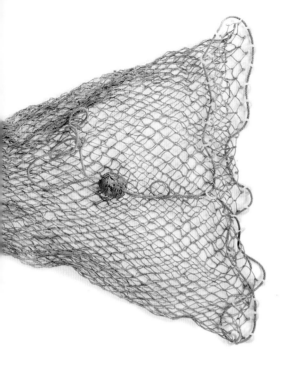

16 Fishing Net (渔网)

What did the early humans do when they felt like eating fish? Did they jump into a pond and grab it or did they throw a spear in the water to stab it? They did both! Eventually, they figured that they could catch many fish together using a fishing net. So, they made a net using spruce root fibres, wild grass, stones and weights. The oldest fishing net, made of willows, dates back to 8300 BC and is called the net of Antrea.

Rock carvings of Alta, which date back to 4200 BC, hint at the usage of fishing nets during the Bronze Age. In 3000 BC, fishing nets were mentioned in ancient Greek literature.

1. According to the passage, what did the early humans use to make a net?
2. Besides the fishing nets mentioned in the passage, what else do people use to catch fish?

17 Pottery (陶艺)

The first kind of vessels used by humans was probably some kind of basket made from reeds. These vessels had one major flaw; they could not be used to hold water. To fix this flaw, ancient humans lined their baskets with clay soil.

Once water was drawn with the baskets, they were left aside. The leftover water would get soaked into the clay and dry up, causing it to harden and shrink, giving it the shape of the basket. Furthermore, if these vessels were dried in the hot sand or sun, they would harden even further. This is how the first clay pots were born.

1. According to the passage, why did ancient humans line their baskets with clay soil?
2. What did humans do to harden the first clay pots?

18 Language（语言）

We would not usually think of language as something that was formally invented. The truth is that nobody knows exactly how humans began speaking in different languages and it is nearly impossible to find out.

Language can be in two forms — oral and written. Written language is a little easier to trace back. Tally marks are said to be the first form of writing. They were used to keep a count of stock, materials and even the passage of time.

The cuneiform script, one of the earliest forms of writing, was developed in Sumer, which lies in modern-day Iraq. It was written by pressing reeds or styluses into clay tablets to make characters. It originated in the third millennium BC. But by the second century AD, it had become obsolete. During the 18th century, interest in the language was renewed and old texts were deciphered.

1. According to the passage, how many forms can language be in? What are they?
2. Did all the languages evolve naturally? Give your reasons.

Meanwhile, Chinese is the single-most spoken language in the world today. The written origins of the language have been traced back to 1250 BC in the late Shang Dynasty, making it one of the oldest surviving languages in the world.

Although most languages evolved naturally, a few attempts were made to invent languages. The most successfully invented language is Esperanto, which originated in 1887.

19 Fermentation（发酵）

Fermentation is a process which involves breaking down substances to create something new. For instance, yogurt, wine, beer and almost every alcoholic beverage is made by fermenting raw materials like milk, potatoes, grapes, etc. Fermenting materials to make alcoholic beverages started in 7000 BC in the Neolithic Chinese village of Jiahu①. The earliest form of alcohol was made by fermenting fruit, rice and honey. In 6000 BC, winemaking gained popularity in Georgia, Caucasus.

The first person to associate yeast with the process of fermentation was French chemist Louis Pasteur. He was a zymologist who defined fermentation as "respiration without air".

> 1. According to the passage, what is the process of fermentation?
> 2. Apart from the food mentioned in the passage, name at least two other kinds of fermented food.

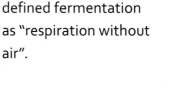

20 Pillow（枕头）

The Mesopotamians were the first to start using pillows around 7000 BC. While we like our pillows to be soft and fluffy, Mesopotamians preferred their pillows to be hard and tough. In fact, early pillows were carved out of stones. Researchers believe that people from the Neolithic Age did not use pillows to sleep comfortably but to keep insects from crawling into their mouths, ears or noses!

> 1. According to the passage, what did people from the Neolithic Age use pillows to do?
> 2. List at least two other materials that can be used in pillows.

The Chinese, too, began using wooden pillows to keep their heads off the ground. However, unlike Mesopotamian pillows, the Chinese had elaborately decorated wooden pillows. A softer version of these pillows, which was stuffed with straw, reeds or feathers, was used by the ancient Greeks and Romans.

---

① Jiahu 贾湖（黄河流域新石器时代遗址）

2.6 million years ago – 3000 BC
（260万年前—公元前3000年）

21 Mortar（砂浆）

Mortar is a thick paste used to stick construction materials together. It is used to hold stones, bricks and other construction blocks together. Plaster of Paris① is also a type of mortar and was the first to be discovered. It was called the gypsum mortar by the Egyptians and was used in the construction of the Egyptian pyramids.

Researchers have found evidence of mortar dating back to 6500 BC. This evidence was found in the Mehrgarh② region of Baluchistan, Pakistan, and the mortar was a mixture of mud and clay instead of stone, mainly because of the abundance of clay during that time.

1. According to the passage, where have researchers found the evidence of mortar?
2. What's the advantage of mortar?

22 Refined Salt（精制盐）

We already know that salt makes our meal tasty. But it does a lot more! Some countries have fought over it and some have even used salt as their currency. Salt has been manufactured and used since 6050 BC during the Neolithic era in Romania. Researchers have found evidence of salty spring water, which was boiled to extract the salt. It is also believed that salt was directly responsible for the immense and immediate growth of that area.

1. According to the passage, how long has salt been manufactured and used in Romania?
2. Besides making meals tasty, what else can people use salt to do?

Salt was used to barter goods in Neolithic times. In fact, slabs of rock salt were used as coins in Abyssinia which is today's Ethiopia. Many years later, Venice fought with Genoa over salt.

① Plaster of Paris 熟石膏　② Mehrgarh 梅赫尔格尔（巴基斯坦的考古遗址）

23 Axe（斧）

The axe is one of the oldest tools used by humans. However, it looked very different from the axes that we see today. The first axe looked like a sharpened stone. It did not have a handle and was therefore called a hand axe. Axes with handles appeared only around 6000 BC. They were attached to a piece of wood and a length of animal hide was wound around them to hold the axe together.

With the dawn of the Bronze Age, humans began to create axes using metals like bronze and copper. Soon, they began making moulds. This helped to replicate and produce axes on a large scale.

Today, we use axes for a variety of purposes, including carpentry, metallurgy and farming.

1. According to the passage, why was the first axe called a hand axe?
2. At different times, humans use axes for different purposes. Give some examples.

24 Irrigation（灌溉）

The concept of irrigation is said to have originated from the river valley civilisations, especially those of Egypt and Mesopotamia. The people would channel the water that overflowed from the Nile and Tigris rivers to their fields. The Egyptians devised a method to measure the level of water, called nilometers. This would help them predict when the river would flood.

Nilometers allowed the Egyptians to prepare for floods. They would divert flood water to lakes via canals and dams. One such canal measured around 20 km!

1. According to the passage, what did the Egyptians use nilometers to do?
2. Besides nilometers, what else did humans use to predict natural disasters? Give one example.

25 Leather（皮革）

Leather is considered to be one of the most important discoveries by humans. Egyptian tombs dating back to almost 3,300 years indicate that sandals, clothes, gloves, buckets, shrouds, bottles and military equipment were made of leather.

Early humans used animal hides to protect themselves from forces of nature such as heat, wind, rain and cold. But this became problematic, as the hides would rot in the heat and stiffen in the cold weather.

The initial process of leather tanning probably started out as a mistake. Early humans probably discovered that drying skins out in the sun made them durable and flexible. Rubbing salt into skins, exposing them to smoke and rubbing animal fat on them were some common methods used to create leather. The art of tanning leather was once a closely guarded secret in primitive times, passing only from father to son!

The ancient Greeks contributed a lot towards developing a tanning formula for leather. They began using water and tree barks to preserve leather.

The British started manufacturing leather on a full-fledged basis when they were introduced to it by the Romans. Eventually, the tanning of leather gained popularity and tanneries were set up all over Britain. Some of these medieval tanneries can be seen even today at Tanner Street, Baker Street and Leather Lane in London.

1. According to the passage, how did the ancient Greeks preserve leather?
2. List two advantages of leather.

26　Glue（胶水）

While the glue that we use today is made from chemicals, back in 4000 BC, it was made from tree sap. Archaeologists have unearthed evidence which suggests that ancient Greeks used adhesives for carpentry. In fact, the Greeks made adhesives from egg whites, blood, bones, milk, cheese, vegetable oils and different types of grains. On the other hand, Romans used tar and beeswax to make glue.

A path-breaking discovery in the adhesive world was that of "Super Glue". An American inventor, Dr Harry Coover, rejected a substance called cyanoacrylate① because he found it too sticky and useless for his research in 1942. A decade later, in 1958, Dr Coover realised that cyanoacrylate was not useless at all! It was then packaged and marketed as Super Glue.

> 1. According to the passage, back in 4000 BC, what was glue made from?
> 2. What is glue used for in our daily life?

27　Papyrus（纸莎草）

The discovery of papyrus and its use as a writing material was perhaps one of the most important inventions in the history of mankind. The Egyptians were the first to feel the need of a lighter writing medium after developing the written language. After inscribing on stones for a long time, the Egyptians started using the sap of the papyrus tree to make writing sheets as early as 2500 BC.

Papyrus sheets remained in use till the 11th century AD. The sheets were rolled or made into scrolls for long documents. Scrolls were made of 20 or more sheets whose ends were stuck together. Loose sheets were hardly ever sold. Eventually, books were made from papyrus and called codex. Papyrus, too, had different qualities. Thicker papyrus was used for packaging, while the best quality and finer sheets were used to write religious or literary scriptures.

> 1. According to the passage, when did the Egyptians start using the sap of the papyrus tree to make writing sheets?
> 2. What writing materials do you know in the history of China? Give at least two examples.

① cyanoacrylate 氰基丙烯酸盐

2.6 million years ago – 3000 BC
（260万年前—公元前3000年）

28 Shoes（鞋）

1. According to the passage, why did early humans have thick toe bones?
2. Do you think shoes are important for us? Why?

The oldest footwear unearthed by archaeologists is a pair of sandals from around 10,000 years ago. The oldest shoes are a leather pair that dates back to 3500 BC. However, humans have been wearing shoes for much longer, almost for 40,000 years! As these shoes were made from perishable materials, no evidence remains.

Our bones grow and change according to our needs. Early humans had thick toe bones that would help them walk and climb through rough terrain. Around 40,000 years ago, these toe bones grew small and weak. This is because shoes take the stress away from the toes. This observation helps us understand the historical timeline of shoes!

29 Sundial（日晷）

The first attempt to record time was made by the Egyptians 3,500 years ago. This ancient clock was called a sundial. A sundial can tell us the time by using shadows to record the position of the Sun in the sky. Also, the first documented evidence of sundial in China is around 574 AD.

A horizontal sundial is very common and derives its name from the way it is made. The most common horizontal sundial uses a surface that is marked with lines indicating the hour of the day. The Sun casts its shadow on the style — a sharp, straight, thin rod — forming a shadow on the surface of the sundial. Depending on the time of the day, the shadow moves to different hour lines that are indicated on the sundial.

1. According to the passage, what is the surface of the most common horizontal sundial like?
2. Do you think the sundial is a great invention? Give your reasons.

30 Wheel（轮子）

The wheel seems like a simple contraption, doesn't it? But it is more complex than what meets the eye. In order to move a heavy vehicle on wheels, the wheels would need to be smooth and equal in size. Their centres would have to be perfectly aligned for the axle to pass through them. The axle would have to fit snugly within the centres of the wheels and yet leave enough space for the wheels to move smoothly.

The first wheel was not used for transportation; it was used for pottery. An updated version of a potter's wheel exists even today. It was only around 300 years after the invention of the potter's wheel that the wheel came to be used for transportation.

The first wheels were heavy as they were formed of a solid, round discus. As a solution to this problem, spoked wheels were invented. These wheels revolutionised transportation and remained in use with minor modifications until the 19th century.

The invention of wire-spoked wheels was a significant achievement, as it made the wheels even more lightweight. The tyre is another achievement that was invented in the late 19th century. It allows the wheel to move more smoothly as compared to its earlier versions.

31 Belt（皮带）

1. According to the passage, why were first wheels heavy?
2. Do you think the wheel is a useful invention? Why?

Today, belts are primarily used for two reasons: to keep one's trousers or skirt snug around the waist and as a fashion statement. But the first belt was worn for practical purposes, such as holding tools.

For a long time, belts were commonly worn by men. They started being considered as a fashion accessory for women only in the 17th century.

The modern belt has many shapes and sizes — thin, broad, braided, studded or jewelled. Even superheroes cannot get enough of the belt — Batman's utility belt is one of the most iconic components of his costume!

1. According to the passage, when did belts start being considered as a fashion accessory for women?
2. List at least two uses of belts.

 **32 Ship（船）**

The first ship was invented around 3000 BC. It was built by tying planks of wood together and stuffing the gaps with dry grass, reeds, and animal hide to make the ship waterproof.

The Egyptians were the first shipbuilders. They made use of sails while travelling downwind and oars while travelling upwind. The Phoenicians built some of the very first warships. Since it was important for the ships to move quickly, they needed several oarsmen to row at once. To serve this purpose, they created a ship called "the bireme" in which the oarsmen sat on two levels, one on top of the other. The Chinese, on the other hand, innovated strong hulls and multiple sails.

The modern ship may not be powered by oarsmen or made from wood, but it still uses elements of all its older counterparts from thousands of years ago.

1. According to the passage, how was the first ship built?
2. What did the Phoenicians do to make the warships move quickly?

33 Saw（锯子）

A saw is a knife-like instrument used to cut through hard materials like wood. It has "teeth" on its edges that help it cut better. Greek mythology states that an inventor called Talos created the saw after observing the jaws of a fish. However, modern archaeology suggests that it was early humans who first came up with the saw. In China, people believe that a Chinese called Lu Ban invented the saw after noticing a kind of grass with zigzag edges.

Consequently, the very first saws were simply sharp objects with jagged edges. This could be anything from a piece of obsidian① to a broken seashell and even a shark's tooth! The ancient Egyptians made the first metal saws from copper. Until the mid-1800s, saws were made by hand. Today, they are made in factories.

> 1. According to the passage, why does the saw have "teeth" on its edges?
> 2. List at least two objects with sharp edges.

34 Cement（水泥）

Cement is a binding substance that is used in construction. The earliest substance used for this purpose was clay. In Egypt, gypsum was used for the same purpose. The Greeks and Romans used lime② that comes from limestone③.

There was one big flaw in this cement, however — it needed to be in a dry environment in order to set properly. This was a huge hurdle in the way of underwater or waterside constructions. It was the Romans who discovered the first recipe for hydraulic cement. They achieved it by mixing volcanic ash or crushed brick into the cement mix.

> 1. According to the passage, what binding substances have been used in construction?
> 2. What was the first recipe for hydraulic cement?

---

① obsidian 黑曜石　② lime 生石灰　③ limestone 石灰岩

2.6 million years ago – 3000 BC
（260万年前—公元前3000年）

35 Bronze（青铜）

Bronze was such an important discovery for humankind that an entire era of history is called the Bronze Age. It is the first alloy to be invented by man. Originally, bronze was made by melting copper and arsenic① together. By the third millennium BC, tin replaced arsenic in the formula and a new form of bronze was invented.

A big obstacle with bronze was that copper and tin② could rarely be found underground together. Thus, the use of bronze involved a lot of trade. In comparison, iron was easily available and therefore more common. That is why the Bronze Age eventually faded away and paved the path for the Iron Age.

1. According to the passage, how was bronze originally made?
2. Why was the Bronze Age finally faded away?

36 Button（纽扣）

The first button was made of a curved shell thousands of years ago in the Indus Valley. Following this, buttons were also found in Bronze Age sites in China (2000 BC–1500 BC) and Ancient Rome. Buttonholes were not invented until the 13th century AD. The original buttons were simply fitted into loops.

The functions of the button varied a lot. The Chinese button was invented for fastening clothing at the very beginning. However, in some European countries, the button was not made to hold two pieces of fabric together innitially. It was used as a fashion symbol and a sign of wealth. In fact, it was said that people could pay their debts simply by plucking an expensive button off their clothes! Later, as more fitting styles came into fashion, buttons helped men and women achieve the right look by fastening their clothing tightly.

1. According to the passage, when were buttonholes invented?
2. How do you like buttons? Why?

① arsenic 砷　② tin 锡

37 Silk（丝绸）

Silk is a natural fibre made by dissolving silkworm cocoons in boiling water. A cloth made from silk has a shimmering effect. This is because of the triangular structure of the molecules in silk, which reflects light from several different angles.

Legend has it that silk was accidentally discovered by a Chinese queen. One afternoon, she sat under a mulberry tree sipping some tea. Suddenly, a silkworm cocoon fell into her cup. When the queen tried to pull it out, it unravelled into thin, shiny threads in her hands.

The Chinese were quick to realise that their discovery was significant. Traders from far and wide traded silk with other countries for all kinds of riches. In fact, silk was so in-demand that its constant trading led to the popularisation of the Silk Route from the East to the West.

Some Chinese immigrants carried silk to Korea. Some say that a Chinese princess married a prince of another country and carried cocoons in her headgear. In this way, the method of creating silk was introduced to other Asian countries and later around the world.

1. According to the passage, why does a cloth made from silk have a shimmering effect?
2. Have you ever heard of the Silk Route? What is it?

38 Wig（假发）

The Egyptians were the first to wear wigs. Most people in ancient Egypt preferred to shave their heads as the climate was extremely hot. However, bald heads were not considered to be attractive. Wigs served a dual purpose of making the heads look like they had hair while also protecting the scalp from the heat.

There were distinct differences between the wigs for upper classes and those for lower classes. The richer sections of society had elaborate wigs that were adorned with silver and gold. These wigs were made of human or animal hair, palm leaves and even wool!

1. According to the passage, what was the dual purpose of the ancient Egyptian wigs?
2. Do you think wigs will be less popular in the future? Why?

39 Pen（钢笔）

The first pens were thin twigs which men used to scratch on clay tablets. Pens that could hold liquid ink were made from hollow reeds or straws. Around 500 BC, the first quills were created. These were made by sharpening the ends of bird feathers to make writing nibs.

Quills were widely used for a long time until the first metal nibs were created in the 19th century. The first fountain pen was created in the late 1800s. The main advantage of this pen was that it could hold a large amount of ink, thus eliminating the need to constantly refill it.

1. According to the passage, how were the first quills made?
2. How many kinds of pens are mentioned in the passage? What are they?

Today, apart from fountain pens, there are several, widely used varieties such as ballpoint pens, roller ball pens, felt pens and pens with ceramic nibs.

40　Kohl（眼线）

Kohl is a form of eye makeup. It is not just used as a cosmetic; it is a coolant and is also believed to protect children from evil eyes. Kohl was historically worn in Egypt, South Asia, the Middle East and other parts of Africa.

The Egyptian recipe for kohl consisted of stibnite[①], which is a sulphide[②] of antimony[③]. One way to make kohl was the dip-dry method, in which a small square of muslin was repeatedly dipped into sandalwood paste and left to dry in the sun, forming a wick. This wick was used to light an oil lamp. The soot thus formed was used as kohl.

Kohl gained popularity in the west only in the 1950s, when women began to apply makeup more liberally than before.

1. According to the passage, what is kohl used as?
2. Why did kohl gain popularity in the west only in the 1950s?

41　Shadoof（桔槔）

The shadoof, was an early crane-like tool with a lever mechanism, used in irrigation since around 3000 BC by the Mesopotamians, 2000 BC by the ancient Egyptians, and later by the Minoans, Chinese (1600 BC), and others. It looked like a long pole with a bucket or a bag attached to the end of it.

The shadoof was important to the ancient Egyptians because it helped water crops. The Nile flooded every June but the Egyptians needed to survive the rest of the year too. Therefore, they created the shadoof to refill the irrigation channels that they had built for the annual flooding.

1. What did the shadoof look like?
2. How did the Egyptians survive the rest of the year except June?

---

① stibnite 辉锑矿　② sulphide 硫化物　③ antimony 锑

2.6 million years ago – 3000 BC
(260万年前—公元前3000年)

42 Iron（铁）

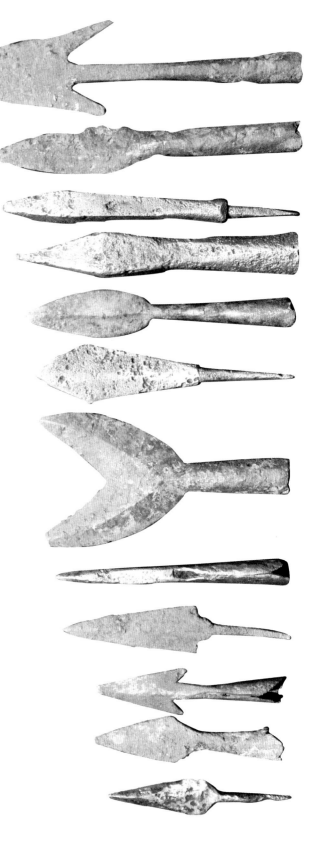

As iron is available in abundance, it is not exactly an invention. However, this versatile metal has been widely used throughout the ages and its use for various purposes could be called a discovery in itself.

The Iron Age was an important era in history when humans made great progress in building tools and implements. Today, it's safe to say that the world would undergo greater suffering if we ran out of iron, instead of gold or silver.

Iron was probably first discovered when early humans tried to burn the ore and found that it melted at a high temperature. Soon, they began replacing bronze with iron in instruments and implements.

It is widely believed that the Asians first came up with the concept of iron smelting, slowly spreading their methods across the world. A common place for smelting iron was a bloomery. Here, the iron was heated until it became a spongy mass. This mass was then hammered into the desired shape.

Over the years, improvements in furnaces and processes led to the discovery of different types of iron, like cast iron[①] and pig iron[②].

1. According to the passage, why was the Iron Age an important era in history?
2. Why does the author say "The world would undergo greater suffering if we ran out of iron."?

---

① cast iron 铸铁　② pig iron 生铁；铸铁

43 Toothpaste（牙膏）

It is believed that the Egyptians began using a paste to clean their teeth around 5000 BC. Ancient Greeks and Romans are known to have used toothpastes, and people in China and India first used it around 500 BC. Ancient toothpastes were used for keeping teeth and gums clean, whitening teeth, and freshening breath. The ingredients of ancient toothpastes differ from the ones that we use today. Back then, ingredients like ash and burnt eggshells were combined with pumice to make toothpaste! The Greeks and Romans favoured more abrasiveness, and used crushed bones and oyster shells. The Romans added more flavouring for good breath as well as powdered charcoal and bark. The Chinese used a wide variety of substances such as ginseng, herbal mints and salt.

Modern varieties of toothpaste gained popularity in the 1800s. The early versions contained soap. Chalk was included in the 1850s. Betel nut was included during the 1800s. In 1873, Colgate mass produced the first toothpaste in a jar. In 1892, Dr Washington Sheffield created the first toothpaste in a collapsible tube. Currently, the use of fluoride has increased in toothpastes as it has been found to reduce cavities.

1. According to the passage, what were ancient toothpastes used for?
2. What does the passage mainly tell us?

2.6 million years ago – 3000 BC
（260万年前—公元前3000年）

44 Weighing Scale（天平）

Weighing scales are said to have originated during prehistoric times. It is believed that the first weighing scales were used in the Mesopotamian civilisation in 4000 BC. For the purpose of bartering, a weighing scale was needed. The first weighing scale of the modern era was designed and built by Leonardo Da Vinci during the late 15th century. In 1897, the first scale with indicators was built. The indicating scales appeared in 1940.

Soon after, the electronic weighing scale was designed and perfected, which continues to rule the market till today.

1. According to the passage, who designed and built the first weighing scale of the modern era?
2. Of all the weighing scales mentioned in the passage, which one do you think is the most accurate? Why?

45 Ploughshare（犁铧）

A ploughshare is a part of the plough. It is the cutting or leading edge of a mouldboard that closely follows the "coulter" or the ground-breaking spikes when ploughing.

Triangular-shaped stone ploughshares were discovered at the sites of Majiabang culture around Lake Taihu. These ploughshares date back to 3500 BC. Ploughshares have also been discovered at the nearby Liangzhu[①] and Maqiao[②] sites that date back to around the same period.

The ploughshare is often a hardened blade dressed into an integral mouldboard. This is done by a blacksmith.

1. According to the passage, where were triangular-shaped stone ploughshares discovered?
2. What does the passage mainly tell us?

① Liangzhu (site) 良渚遗址 ② Maqiao (site) 马桥遗址

46 Dentistry（牙科）

The first known case of treating tooth-related problems dates back to 7000 BC, where the Indus Valley civilisation shows evidence of treating tooth decays. The first method used for treating tooth decay involved bow drills. These tools were used for wood works as well as treating tooth problems!

> 1. According to the passage, when did dentistry take a scientific turn?
> 2. How can we protect our teeth? List at least two ways.

The Sumerians thought that worms made small holes in teeth and hid inside them. The idea of a worm residing in teeth and causing dental pain actually lasted until the 1700s!

In ancient Greece, Hippocrates and Aristotle wrote about treating decayed teeth and extracting them to keep the pain away. During this time, extraction was done using forceps, which were used to treat several diseases during the Middle Ages.

During the 18th century, dentistry took a scientific turn. It was revolutionised during the late 18th and 19th centuries. For many centuries, rich patients would get gold fillings in their teeth as a symbol of wealth. False teeth made from porcelain were invented in 1770. Amalgam① was first used in Europe around 1820.

The dentist's chair was invented in 1790 by Josiah Flagg.

① amalgam 汞合金，汞剂

2.6 million years ago – 3000 BC
(260万年前—公元前3000年)

47 Pants (裤子)

Archaeologists Ulrike Beck and Mayke Wagner excavated two ancient graves in a cemetery in Xinjiang, China. Among the remains, they discovered two pairs of well-preserved, woollen pants. Research shows that these pants were between 3,000 and 3,300 years old, making them the oldest-known pairs of trousers to be discovered.

This time period corresponds with the rise of "mobile pastoralism" in Central Asia, where nomads began moving their herds across the land on horseback. Tunics and robes were not comfortable or suitable for long, bumpy rides as well as for battles. As a result, pants were created.

1. How old were these pants discovered in ancient graves in a cemetery in Xinjiang, China?
2. According to the passage, why were pants created?

48 Paved Road (铺面道路)

Around 2500 BC, a road built in Egypt by Pharaoh Cheops is believed to be the earliest paved road on record. It is 1,000 yards long and 60 feet wide. It led to the site of the Great Pyramid. Since it was used only for this job and never for travel, Cheops's road was not, in a true sense, a road like the later trade routes, royal highways and impressively paved Roman roads.

Those who built roads during the late 1800s depended solely on stone, gravel and sand for construction. Water would be used as a binder to give some unity to the road surface.

1. According to the passage, how long and how wide is the earliest paved road on record?
2. Do you know any paved roads with a long history in the city you live in?

2999 BC - 1 BC
（公元前2999年-公元前1年）

49 - Sewage System（下水系统）

Even though it may be awkward to talk about, all of us have to answer nature's call. Today, most civilised people use indoor lavatories. Have you ever wondered what happens after you flush? Where does all the sewage go?

In order for you to have the luxury of a toilet within your own home, a system of pipes and channels need to be in place first. The first sewage systems were not quite systems at all. They were toilet-like cavities that drained to channels just outside.

The first instance of a sophisticated sewage system comes from the Indus Valley civilisation which flourished around 3000 BC. Each house had a bathroom and latrine. These were connected to sewers that led to the river.

One of the greatest wonders of ancient architecture is the Great Bath in Mohenjo-Daro. It was a public bath, not unlike the swimming pools of today. It had a hole at one end which was used to drain the water out.

> 1. According to the passage, what were the first sewage systems like?
> 2. What's the difference between the Great Bath in Mohenjo-Daro and the swimming pools of today's?

50 Concrete（混凝土）

The word "concrete" is derived from the Latin word "concretus", which means condensed or compact. Concrete is a mixture of water, cement and certain other materials. It is normally strengthened by using rods or steel mesh before it is poured into moulds to form blocks. Interestingly, the history of concrete dates back to Rome around 2,000 years ago. Concrete was essentially used in the construction of aqueducts and roadways in Rome.

Romans are known to have used concrete for building roads on a large scale. Interestingly, they built approximately 8,500 km of roads using concrete. Concrete is a very strong building material. Historical evidence shows that Romans used several materials such as quicklime, pozzolana and an aggregate of pumice, substances like animal fat, milk and blood, as mixtures for building concrete.

John Smeaton, an engineer, made concrete by mixing coarse aggregate of pebbles and powdered brick, and added this mixture to cement. This is the mixture that we continue to use even today. From 1756 to 1759, he built the third Eddystone Lighthouse in Cornwall, England. It was during the planning of this lighthouse that Smeaton invented the hydraulic cement. Another major development occurred in 1824. An English inventor, Joseph Aspdin, invented Portland cement by burning grounded chalk and finely crushed clay in a limekiln until the carbon dioxide evaporated, resulting in strong cement. The first systematic testing of concrete took place in Germany in 1836.

1. According to the passage, what is concrete?
2. Give two examples of major development of concrete.

51 Plier（钳子）

No one knows for certain when pliers were invented. They were probably invented out of necessity — humans felt the need of tongs to help them work with hot fires. With the dawn of the Metal Age, hot fires turned into blazing hot furnaces. Makeshift wooden pliers would no longer do. Thus, the first bronze pliers were made in circa 3000 BC.

Today, we have pliers of varying sizes that are used for various purposes. There are surgical pliers, dental pliers, laboratory pliers, kitchen pliers and even jewellery pliers!

1. According to the passage, why were pliers probably invented?
2. What kind of pliers have you known? Name at least two kinds.

52 Dam（水坝）

Like many other ancient inventions, the Egyptians built the first dam. This was a simple dam made of rock and gravel which resisted the force of water simply with its weight. However, this dam was not very effective as water eventually managed to trickle through the rock.

The Mesopotamians had better luck with their dam, which was made of soil and clay. Wooden dams are also believed to have been constructed.

The Chinese ealiest dam was constructed in 453 BC, which was called Zhibo Qu. Around 256 BC, a Chinese official, Li Bing, built Feishayan (Flying Sand Weir) in Sichuan province. This dam helped control the amount of water and deposit sand and stones from the upstream.

The Romans built the first concrete dam in 100 BC, but it was only in the 17th century that the Spanish perfected the art of making dams. They travelled around the world, taking their building methods with them and sharing their knowledge with the rest of the world.

1. According to the passage, who built the first dam?
2. Do you know any other dams in China?

2999 BC – 1 BC
（公元前2999年—公元前1年）

53 Soap（肥皂）

Soap has been used since 2800 BC and was invented by the Babylonians. Back then, it was simply a cake made from animal fat. Do you ever wonder how humans came to the realisation that rubbing themselves with animal fat would make them cleaner? Legend says that animal sacrifices might be the reason. People noticed that fat ran into the river water, forming a clay-like substance that left their clothes cleaner when they washed them with that water.

1. According to the passage, how long has soap been used?
2. Besides animal fat, what else can you find in soap?

Even though soap is such an ancient invention, liquid soap and shower gels did not come into existence until about a hundred years ago.

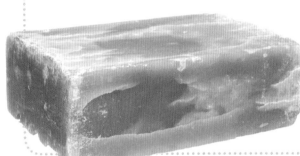

54 Toilet（厕所）

The first toilets were nothing more than holes in the ground. However, toilets with a flushing system are not a recent invention. The first such toilets were found in the ancient Minoan, Egyption and Harappan civilisations.

The first instance of the modern flush toilet that we know today appeared in the 16th century. It was designed by Sir John Harington, who installed one for Queen Elizabeth I. However, she refused to use it as she found it too loud!

As the population increased, so did the production of sewage. This was when the toilet gained popularity.

1. According to the passage, what were the first toilets like?
2. What is the passage mainly about?

55 Dye（染料）

Until the 19th century, all dyes were made from natural substances. Colours in the red, brown and orange families, which are easily found in nature, were the very first dyes to be used by humankind. The process for creating these dyes was simple — the source of the dye, along with the fabric to be dyed, would be boiled in water until the colour transferred from the source to the fabric.

Colours like blue were also developed with the help of flowering plants. Indigo was one of the most widely produced and sought-after natural dyes. However, it wasn't until much later, in 1856, that the first man-made dye was discovered.

1. According to the passage, what colours are easily found in nature?
2. What do you think of the process for creating the dyes in the past?

56 Column（柱子）

A column or pillar is an architectural structure that distributes the weight of a heavy structure onto the ground below it. This allows the use of ceilings without the need of actual solid walls to support it. There are instances of columns found in ancient architecture, including that of the Egyptian, Assyrian and Minoan civilisations. However, it was the Persians who built some of the most ornate and beautiful columns.

During the Middle Ages, the use of columns diminished greatly. They reappeared in Renaissance architecture in the 15th century.

1. According to the passage, what is a column?
2. Where can we find the columns in ancient architecture in China?

2999 BC – 1 BC
（公元前2999年-公元前1年）

57 Baking（烘焙）

Baking was one of the earliest forms of cooking. Ancient humans would crush grains and mix them with water to form a paste. This paste would be laid on a hot rock, which would cook it into some kind of flat bread. The Romans were known to be very keen bakers. They had lavish parties and their spreads would include new and innovative baked foods.

During the Middle Ages, consuming pastries and baked products became a sign of prestige. The rich ate bread made from refined flour, whereas the poor had to settle for coarse bread. It was only during the Industrial Revolution of the 18th century that baked goods became available just as easily to the common folk.

> 1. According to the passage, when did baked goods become available just as easily to the common folk?
> 2. What other forms of cooking do you know besides baking? Name at least two forms.

58 Frying（油炸）

While frying is an ancient invention, it was only practised by a few civilisations, like the Egyptians and Mesopotamians. This is because other civilisations had not discovered methods by which they could crush the seeds to extract oils. It is believed that fried cakes were eaten in Egypt since 2500 BC!

> 1. According to the passage, why was frying only practised by a few civilisations?
> 2. When the Europeans settled in the Americas, did the natives accept the idea of frying their food? Why?

Isn't it a strange idea that cooking something in oil and fat would make the food tasty? Native Americans thought so, too! When the Europeans settled in the Americas, the natives were repulsed at the idea of frying their food. However, the method found its way into their cuisine. Gradually, people began frying meats and the method soon became popular in cuisines around the world.

59 Plough（犁）

The practice of irrigation and the systematic plantation of crops began a long time ago. The first evidence of planned farming dates back to the Mesopotamian era. The plough is a very important tool in farming; its earliest evidence dates back to 2000 BC. Interestingly, early British Law directed ploughmen to use only those ploughs that were constructed by them.

Simple ploughs were developed from handheld hoes created by the Egyptians. These ploughs were pulled by oxen, camels and elephants. However, some of these methods have been frowned upon by historians because they involved cruelty towards animals. Till around the 17th century, farmers would tie the ploughs to the tail and horns of the animals.

1. According to the passage, when did the earliest plough appear?
2. What do you think of the ploughs pulled by oxen, camels and elephants?

60 Perfume（香水）

The art of making perfume began in the Mesopotamian era around 2000 BC and was continued with much fervour by the Romans and the Persians. The word perfume has Latin origins and comes from the word "perfumare" which means "through smoke".

According to historians, a Mesopotamian woman called Tapputi distilled flowers, calamus and oil to make perfumes in 2000 BC. The process involved filtering and mixing these ingredients with other aromatics and setting them aside for a long period of time.

The world's oldest perfumes were discovered in 2005 by archaeologists in Cyprus. These perfumes are older than 4,000 years. Evidence suggests that they had been manufactured in a 43,000 square feet

perfume factory. These perfumes were made from flower extracts mixed with spices like almond, bergamot and myrtle.

1. According to the passage, when did the art of making perfume begin?
2. How did Tapputi make perfumes in 2000 BC?

2999 BC – 1 BC
（公元前2999年—公元前1年）

61 Architectural Arch（拱门）

The earliest evidence of architectural arches used for gates dates back to the Bronze Age. This evidence was unearthed in the city of Ashkelon, (now in modern day Israel), in 1850. True arches or perfectly round arches were often used for underground buildings like the drainage systems in Mexico, the Levant and ancient Near East.

The Romans were the first to start using arches extensively in their structures. They learnt the art of constructing arches from the Etruscans and, after much improvisation, used the arches for above-the-ground structures like bridges, aqueducts and gates.

The triumphal arch was invented by the Romans and used as a military monument throughout their civilisation.

1. According to the passage, when was the earliest architectural arches used for gates?
2. What did the Romans do after they learnt the art of constructing arches from the Etruscans?

62 Ruler（尺子）

Measurements are crucial to building and construction. Thus, rulers have existed for years. Initially, they were made of ivory and wood. The earliest evidence of a ruler was found in the remains of the Indus Valley civilisation. It dates back to around 2400 BC.

Before we had this instrument, people used their body parts for measuring things. For example, a cubit was the length of a man's arm from the elbow to the tip of his middle finger. However, this system had its drawbacks — everyone's arms are not equal in size! The person who took the measurements would ideally have to build it himself/herself.

Thus, with the invention of the ruler, a common standard unit of measurement was introduced which would be the same for all.

1. According to the passage, what did people use for measuring things before they had rulers?
2. What do you think of rulers? Give your reasons.

63 Umbrella（伞）

Today, we cannot imagine living without fans and air conditioners to keep us cool. They make our lives more bearable. For the people who lived during the ancient times, the umbrella was such an invention. It greatly improved the quality of their lives simply by providing them additional protection from the elements of nature.

The umbrella was first used as a shade against the sunshine. This concept later evolved into the parasol. However, one difference between an umbrella and a parasol is that a parasol is usually held over another person instead of holding it over oneself.

The first umbrellas were symbols of rank and royalty. In ancient Egypt, fair skin was associated with high class. That is why the kings and the highest nobility had their servants hold umbrellas over their heads as they walked. In China, Lu Ban took the credit for inventing the umbrella.

1. According to the passage, how did the umbrella greatly improve the quality of people's lives during the ancient times?
2. What's the passage mainly about?

2999 BC – 1 BC
（公元前2999年—公元前1年）

64 Chariot（马车）

Archaeologists believe that chariots were invented in 2000 BC in the ancient Near East civilisations. More prominent evidence of full-spoke chariots were found from the remains of buried chariots at the Andronovo sites① in the Sintashta-Petrovka Eurasian culture.

It seems the power of Chinese states and dynasties was often measured by the number of chariots they were known to have. The earliest archaeological evidence of chariots in China is of a chariot burial site discovered in 1933 at Hougang②, Anyang in Henan province dating to the rule of King Wu Ding of the late Shang Dynasty (1200 BC). The inscriptions suggest the Shang used them as mobile command-vehicles and in royal hunts.

The Zhou Dynasty made more use of the chariot than did the Shang Dynasty. From the 8th to 5th centuries BC the Chinese use of chariots reached its peak. Although chariots appeared in greater numbers, infantry often defeated charioteers in battle. However, chariot warfare became all but obsolete after the Warring States period (475–221 BC).

1. According to the passage, which dynasty made more use of the chariot, the Zhou Dynasty or the Shang Dynasty?
2. Why did chariots play very important roles in ancient warfare?

65 Scissors（剪刀）

Historians suggest that modern scissors were invented in Rome around 100 AD. However, a primitive pair of scissors was made in Egypt in 1500 BC. It was called the spring scissors because the handles of its bronze blades were connected with flexible, curved, bronze spring. When the handles were squeezed together, the spring ensured that the blades stayed aligned. In fact, Europeans used the spring scissors until the 16th century. The bronze spring scissors were widely used in China, Japan and Korea too.

1. According to the passage, why was a primitive pair of scissors called the spring scissors?
2. Do you think scissors are useful? Give your reasons.

---

① Andronovo sites 安德罗诺沃遗址　② Hougang (site) 后冈遗址

66 Refined Sugar（精制糖）

Refined sugar, as we know it today, refers to the pure white crystals of sugar that we usually mix into our beverages. The sugarcane plant can be crushed to release juice. The people of ancient India discovered that by boiling this juice, it would reduce to form small, rocky particles.

> 1. According to the passage, how did the people of ancient India discover the method of making refined sugar?
> 2. Why do people consume a large amount of sugar every year?

Initially, sugar was so rare that it was called "white gold". But as sugar spread around the whole world, newer and cheaper refining methods were discovered, thus bringing down the cost.

Today, the world consumes a total of about 120 million tonnes of sugar every year. This figure is steadily rising by two million tonnes a year.

67 Steel（钢）

The first evidence of steel manufacturing can be traced back to 2000 BC at an archaeological site in Anatolia. During the same era, steel was also being manufactured in East Africa. The alloy was also used to make weapons like the falcata in the Iberian Peninsula. Noric steel① was used to make artillery for the Roman military. The Chinese, too, used quench-hardened steel② to make weapons in the Warring States period, i.e., from 475 BC to 221 BC. The Han Dynasty produced steel by melting wrought iron③ and cast iron together, which produced the best carbon intermediate steel④ in the first century.

> 1. According to the passage, what was steel mainly used for in the ancient era?
> 2. List at least two uses of steel in the modern society.

It has been speculated that steel was produced in iron smelting factories and bloomeries in the ancient era. The Spartans also produced steel extensively in 650 BC.

① Noric steel 诺里克钢　② quench-hardened steel 淬硬钢　③ wrought iron 熟铁　④ carbon intermediate steel 中炭钢

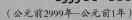

68 Archimedes' Screw（阿基米德螺旋泵）

The Archimedes' screw or the screw pump is an ancient tool used to transfer water from low-lying water bodies to farms. Greek mathematician Archimedes of Syracuse invented the screw in 300 BC.

Apart from irrigation, the Archimedes' screw was used to drain out land covered by sea water in the Netherlands. Here, water was pumped out from a shallow part of land covered by sea. This water was then used for irrigation. The Archimedes' screw was used by the Assyrian King Sennacherib from 706 to 608 BC. Even the classical author, Strabo, describes that screw pumps were used to water the Hanging Gardens①. Screw pumps evolved when German engineer Konrad Kyeser installed the crank mechanism in them.

1. According to the passage, what was the use of the Archimedes' screw?
2. What is the passage mainly about?

① Hanging Gardens 空中花园

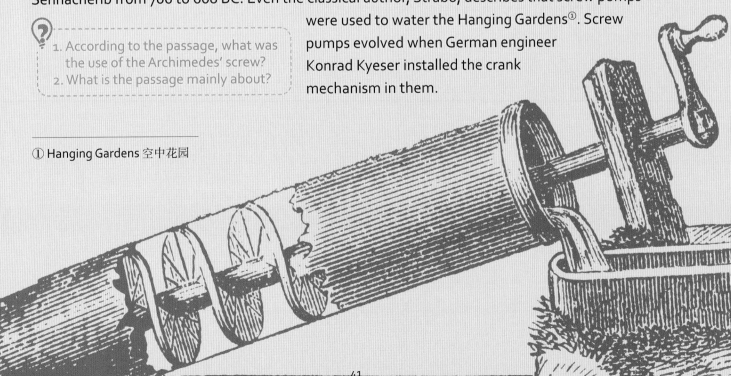

 69 Scarf（围巾）

Rome introduced the world to scarves. Men started using "sudarium" or "sweat cloth" by tying them to their belt or wearing them around their necks in the third century BC to wipe their necks and faces during summer. Eventually, women began using scarves made of wool, pashmina and silk, and made them a fashion statement.

Chinese warriors wore cotton scarves to identify themselves during Chinese Emperor Cheng's rule. Following the example, soldiers in Croatia wore scarves which were decorated as per their ranks. The officers in the Croatian army wore silk scarves while the others wore plain cotton scarves.

1. According to the passage, why did men start using "sudarium" or "sweat cloth"?
2. What are the most common materials used to make fashion scarves?

70 Shower（淋浴）

The Romans believed in bathing several times a week and hence followed the Greeks when it came to making showers. The earliest showers known to humankind were built by the Greeks in 800 BC. The Greeks built large rooms meant for showering which resemble the locker rooms of today. These shower rooms were meant for both the elite and commoners, and were found in the city of Pergamum.

The Romans followed this system and improvised by making lead pipes to both pump in and pump out the water. They also made a complex and intricate sewage system to cater to the sanitation needs of society.

1. According to the passage, who built the earliest showers?
2. Do you like taking a shower or having a bath? Give your reasons.

71 Gloves(手套)

Gloves are considered to be quite unique. Herodotus, in "The History of Herodotus" (440 BC), mentions how Leotychides was incriminated by a glove or "gauntlet" containing silver that he received as a bribe. There are occasional references to the use of gloves among the Romans as well. Pliny the Younger (c. 100 BC), who worked as his uncle's shorthand writer, wore gloves during winter so that his work wouldn't get affected.

During the 13th century, gloves were worn by ladies as a fashion ornament. They were made of linen and silk. It was in the 16th century that gloves grew more popular when Queen Elizabeth I set the fashion for wearing them.

1. According to the passage, why did Pliny the Younger wear gloves during winter?
2. Besides warmth, what else can gloves be used for?

72 Bathtub(浴缸)

According to historians, Romans are known to be the champions of bathing. It was the Romans who first began to bathe every day in public baths. Early plumbing systems provided for baths date back to 3300 BC. However, the evidence of actual bathtubs dates back to 500 BC in Rome.

The first personal bathtub was discovered on the Isle of Crete with a 5-feet-long pedestal made with hardened pottery. The Romans also used marble for tubs, and lead and bronze for pipes, which created a complex sewage system.

1. According to the passage, where was the first personal bathtub discovered?
2. What did the Romans use for tubs and pipes?

73 Lock and Key（锁具）

The earliest lock and key device, dating back to around 4,000 years, was found near Nineveh, the capital of ancient Assyria. The ancient Egyptians were known to use wooden pin locks. These locks had a series of pins inside them. When the right key was inserted, it would push the pins aside, allowing the lock to open.

The first metal locks appeared towards the end of the first century. Metal keys, too, were used. The Romans were known to lock their safes and wear their keys on their fingers as rings. This not only kept them safe, but also allowed them to show off the fact that they had valuables that were locked away.

1. According to the passage, when did the first metal locks appear?
2. Do you think the lock and key device is a helpful invention? Give your reasons.

74 Calliper（测径规）

Calliper is a measuring instrument that resembles a compass. It is very useful for minute measurements. The oldest specimen was found in a Greek shipwreck near Italy. It was made in the sixth century BC. Thus, the Greeks and Romans were the first ones to use callipers.

A bronze calliper from the ninth century AD was found with an inscription that helped to date it. It could measure distances as small as a tenth of an inch.

The modern calliper was invented in 1851 by an American named Joseph R. Brown. It can measure even a thousandth of an inch.

1. According to the passage, what is the calliper?
2. What is the passage mainly about?

2999 BC – 1 BC
(公元前2999年–公元前1年)

75 Aqueduct（高架渠）

Aqueducts refer to the channels or tunnels that carry water from one place to another for easy usage. The river valley civilisations of India, Persia, Egypt, etc., had plumbing systems in place to bring water into people's homes.

However, the aqueduct system of ancient Rome is said to be one of the most sophisticated examples of ancient plumbing. It consisted of channels that carried water from lakes and rivers to homes, public baths and even to farms. These channels were constructed from terracotta, stone, wood or metal.

Over the course of half a century, 11 aqueducts were built in Rome. Some of these transported water from a distance of around 100 km. The aqueducts were built at a slightly downward slope. Thus, gravity was enough to push the water through them.

Most of the aqueducts were built underground. But a 50 km stretch of one particular aqueduct was built over a valley in the form of a stone arch. Many of Rome's aqueducts fell out of use due to a lack of maintenance. Some were destroyed by enemy attacks.

The aqueducts were responsible for Rome's thriving population and the Romans certainly felt their loss. After the aqueducts fell out of use, the Roman population fell from a million to around 30,000.

1. According to the passage, why did Rome's aqueducts fall out of use?
2. Do you think aqueducts were important to the Romans? Give your reasons.

76 Catapult（石弩）

Historically, the catapult was a weapon that used to fling heavy rocks over a great distance. The story of its invention is rather interesting. Dionysius the Elder of Syracuse, Greece, was one of the worst tyrants of the ancient world. He wanted to build a new weapon to strengthen his army. So, he invented the catapult around 400 BC.

Another kind of catapult, the "ballista", was derived from the original design. It was built to shoot arrows instead of heavy loads. The Romans added wheels to the catapult so they could move it around easily.

1. According to the passage, why did Dionysius the Elder of Syracuse, Greece invent the catapult?
2. What's the difference between the "ballista" and the original catapult?

77 Barrel（桶）

In comparison to the inventions of today, the idea of the barrel being a revolutionary invention seems rather ridiculous. However, before the barrel came into the picture, clay pots were used to transport goods. These were not only fragile, but heavy to carry to and fro.

On the other hand, the barrel had a rounded shape, which meant that it could be rolled. Its straight top and bottom meant that it could be stacked easily. It also had handles and in some cases, even wheels.

The rounded sides of the barrel were built using the same methods used to build boats, as discovered by the ancient Egyptians and Phoenicians.

1. According to the passage, what did the barrel look like?
2. What is the passage mainly about?

78 Magnifying Glass（放大镜）

The magnifying glass is said to have been invented in 1250 by an English named Roger Bacon. However, references to similar devices have been found in many instances throughout history.

In 423 BC, Greek playwright Aristophanes wrote a play called "The Clouds" in which magnifying glasses were sold for the purpose of starting fires.

Four hundred odd years later, Pliny the Elder described the same effects with a glass globe filled with water. Both he and Seneca the Younger noted that it could be used to read tiny letters.

1. According to the passage, what could magnifying glass be used for?
2. Do you think a magnifying glass can start fires? Give your reasons.

79 Wheelbarrow（独轮车）

Surprisingly, during ancient times, wheelbarrows were not widely used for agricultural and farming purposes. However, they seemed to be a common instrument for carrying light to medium loads at Greek construction sites.

The oldest clue of the existence of a wheelbarrow comes from an ancient Greek list of building supplies. It describes an item called the "monokyklos", which means "one-wheeler".

However, there is no way to ensure if they were indeed referring to the wheelbarrow. Thus, several historians believe that it was the Chinese who actually invented the wheelbarrow around 100 BC.

1. According to the passage, where does the oldest clue of the existence of a wheelbarrow come from?
2. What do you think of the wheelbarrow? Give your reasons.

Pulley (滑轮)

The pulley is yet another invention that stems from the creation of the wheel. It uses a wheel, axle and a belt or rope to pull heavy loads with little force. The wheel is placed on the axle. It controls the movement and direction of the rope.

The pulley was first mentioned in a Greek text from the fourth century BC. Evidence also states that in 1500 BC, the Mesopotamians used simple rope pulleys to hoist water from wells. However, it is very likely that the pulley existed much earlier, even though there isn't enough evidence to support the theory. The prehistoric monument Stonehenge, is believed to have been built using pulleys.

A pulley that uses one wheel only is called a simple pulley. A compound pulley distributes the weight across several wheels, thus making it possible to haul even heavier weight with considerable ease. Greek inventor Archimedes is said to have invented the compound pulley.

It is said that he had the idea for the compound pulley when he was watching ships being hauled towards the port by several men. With the compound pulley he made, he managed to haul an entire warship with its crew using only the pulleys and his own strength. Just as the famous words he once said: "I could even move Earth if I had a place to stand!"

1. According to the passage, how does the pulley work?
2. What is the passage mainly about?

2999 BC – 1 BC
（公元前2999年–公元前1年）

81 Waterwheel（水车）

The wheel is an iconic invention because of its many applications. One such application was the waterwheel. Initially, it served one of two purposes — irrigation and power generation. It was moved manually with the help of animals or by the water's current.

Waterwheels have grooves on them. When water strikes the wheel, the grooves catch it. The weight of the water in the grooves causes the wheel to move, creating kinetic energy. The wheel then deposits the water into channels for irrigation. The movement of the wheel itself can be used to churn or move machines, like those in mills.

Waterwheels were simultaneously invented in different parts of the world. The Greeks and the Romans first created the waterwheel between the third and first centuries BC. By the first century AD, the Chinese were also using horizontal waterwheels to power their mills.

Even though waterwheels are not widely used today, the modern invention of a hydraulic turbine is heavily derived from it. Water sets the turbine's rotors into motion and the kinetic energy derived from it is used to generate electricity.

1. According to the passage, what was the waterwheel used for?
2. What is the passage mainly about?

82 Plumbing（管道系统）

Around 1700 BC, the Minoan Palace of Knossos① on the Isle of Crete had four separate drainage systems that emptied into great sewers constructed from stone. Terracotta pipes were laid below the palace floor, which could not be seen. Each section was about 2½ feet long, slightly tapered at one end and nearly one inch in diameter. It provided water for fountains and faucets of marble, gold and silver that jetted hot and cold water. This is what we know as the first evidence of plumbing. Concealed in the palace latrine was the world's first flushing "water closet" or toilet, which had a wooden seat and a small reservoir of water. This device was lost for thousands of years. It was during the 16th century that Sir John Harington invented a "washout" closet, which was similar to the earlier one.

1. According to the passage, what is plumbing?
2. How do you like the invention of plumbing? Give your reasons.

83 Compass（指南针）

The first compass was invented in China and it did not look much like the compasses that are in use today. A mineral called lodestone, which is composed largely of iron ore, was found to orient itself in the north-south direction no matter how it was kept.

The first compass had a spoon-shaped pointer which was made of lodestone. The base of this compass was a bronze discus which was inscribed with different constellations. Soon, it was understood that lodestone oriented itself with Earth's poles.

The magnetised needle that we see in compasses today only appeared in the 11th century. The needle could be placed on water to make a wet compass. It could be placed on a pivot to make a dry compass. It could also be hung from a thread.

Compasses made it possible for people to travel further. Sailors from China managed to travel as far as the Middle East without getting lost. Today, magnets have become far more sophisticated. Magnetometers are even embedded in smartphones, which enable them to act as compasses.

1. According to the passage, what is the working principle of the compass?
2. Other than sailors, who else do you think need the compass to assist them? Give your reasons.

① Minoan Palace of Knossos 米诺斯王宫

2999 BC – 1 BC
（公元前2999年—公元前1年）

84 Palanquin（轿子）

A palanquin is a means of transport. It consists of a covered seating area that rests upon one or two horizontal poles. The poles are then hoisted upon the shoulders of the porters, who carry it around from one place to another.

In China, the elite travelled in light bamboo seats, also called litter, supported on a carrier's back like a backpack. Wooden carriages on poles appeared in painted landscape scrolls, in the Northern Wei Dynasty and the Northern and Southern Song Dynasty.

A commoner used a wooden one while the mandarin class litter was enclosed in silk curtains. The bridal chair was of utmost importance. A traditional bride is carried to her wedding ceremony by a "shoulder carriage". This was decked in an auspicious shade of red, richly ornamented and gilded, and was equipped with red silk curtains to screen the bride.

1. According to the passage, how did the litter for a traditional Chinese bride look?
2. What are the advantage(s) and disadvantage(s) of a palanquin?

85 Candle（蜡烛）

Candles were among the earliest inventions, as shown by candlesticks from Egypt and Crete, that date back to at least 3000 BC. Evidence of candles in China can be seen by examining their metal furniture, which had prongs to hold candles. In India, candle wax was made by boiling cinnamon. Other sources of wax included insects, nuts and seeds.

One strange source of wax that was used in ancient times was the eulachon or candlefish. As this fish has very high levels of body fat, it could be dried and used as a candle!

Candles soon grew very popular because they quickly became an essential part of religious ceremonies. As candles burn at a relatively constant rate, they were also used to tell time.

1. How many ingredients were mentioned in the passage to make candle wax? What are they?
2. Other than religious ceremonies, in what other occasions are candles used?

86 Lever（杠杆）

Have you ever sat on a seesaw? If you have, you will know what a lever is. It consists of a beam and a fulcrum. When you apply force on one end, the other end is lifted up. It is easier to push down than to lift up, as when you do the former, the force of gravity also works with you. Thus, levers make the job of lifting easier.

It's impossible to gauge exactly when levers were invented. It is suspected that they were used in ancient Egypt to build pyramids. Greek inventor Archimedes also published laws that helped us understand the lever better.

1. According to the passage, why is it easier to lift objects up with a lever?
2. The author mentioned that levers were used when Egyptians built pyramids. In your opinion, what's levers' job in building pyramids?

87 Dome（圆屋顶）

A dome is a round, architectural structure that usually forms the roof of a building. Simple, dome-like structures have been seen throughout history. Prehistoric structures from 10000 BC used mammoth tusks and bones to create a curved ceiling. An Inuit igloo also has a domed ceiling.

1. According to the passage, what were the features of Roman domed buildings?
2. Why does the author mention the example of an Inuit igloo?

The Romans pioneered large-scale domes. These domes required strong base walls to hold them up. After the fall of the Roman Empire, the eastern Roman or Byzantine Empire carried the legacy forward. The Hagia Sophia① is famous for its huge dome and still remains one of the greatest architectural wonders of the world.

① Hagia Sophia 圣索菲亚大教堂

88 Milling（磨盘）

Milling or grinding is the process of breaking food into finer particles so that it can be consumed easily. Milling also refers to separating, sizing or classifying any kind of material, for example, rock crushing or grinding to produce a uniform size of materials for construction purposes, separation of rock soil for the purpose of land fill or land reclamation activities.

The story behind the invention of milling may never be known to us, for there is no record of where it was first used. The very first type of grinding mill was a mortar and pestle. The grains were powdered by hand in a mortar and pestle. Soon after, milling required a working animal, i.e. a donkey mill, or through a windmill or watermill. Today, mills are also powered by electricity.

1. According to the passage, why do people mill things?
2. What does "particle" mean in paragraph 1 based on the context?

89 Traditional Chinese Calendar（传统中国农历）

The traditional Chinese calendar is lunisolar, that is, it is based on exact astronomical observations of the longitude of the Sun and the phases of the Moon. It is based on the cycle of the Moon as well as on Earth's course around the Sun. A month on this calendar is 28 days long, and a normal year lasts from 353 to 355 days. To keep the calendar in sync with the Sun and the seasons, the Chinese add an extra leap month about once every three years. The beginnings of the Chinese calendar can be traced back to the 14th century BC. According to legend, Emperor Huangdi invented the calendar in 2637 BC.

Although China uses the Gregorian calendar for civil purposes, lunar calendar is still used for determining festivals. Various Chinese communities around the world also use the lunisolar calendar.

1. What's lunisolar calendar according to the passage?
2. After reading the passage, what do you think of the traditional Chinese calendar?

90 Oven（炉子）

An oven is a thermally insulated chamber used to heat, bake or dry a substance and is most commonly used for cooking. The earliest ovens were found in Central Europe, dating back to 2900 BC. They were roasting and boiling pits. The first written historical record of an oven dates back to 1490 in Alsace, France. This oven was made of bricks and tiles.

Inventors began enhancing wood-burning ovens to contain the smoke they produced. Fire chambers were invented to contain the wood fire and holes were built into the top of these chambers so that cooking pots with flat bottoms could be placed directly on them. The 1735 "Castrol stove" or "stew stove" invented by French architect François Cuvilliés completely controlled the fire and had several openings covered by iron plates with holes.

Ovens were used by cultures that lived in the Indus Valley and pre-dynastic Egypt. By 3200 BC, each mud-brick house had an oven in settlements across the Indus Valley. Ovens were used to cook food and make bricks. Pre-dynastic civilisations in Egypt used kilns around 5000–4000 BC for pottery.

1. According to the passage, what are ovens used for?
2. What is the passage mainly about?

2999 BC – 1 BC
（公元前2999年—公元前1年）

91 Postal Service（邮政服务）

The origins of postal systems can be traced back to 2000 BC, in Egypt. During the sixth century BC, the Persian Empire, under the rule of Cyrus the Great, used a system of relay messengers.

In China, a post house service had been started by the Zhou Dynasty. It was mostly used to convey official documents. This system had relays of couriers who changed horses at relay posts that were 14.5 km apart. The system flourished under the Han Empire from 202 BC to 220 AD, when the Chinese encountered the Romans and their postal system. The Roman system, known as "curcus publicus" was the most highly developed system in the ancient world. Their messengers were known to cover a distance of almost 270 km over a day and night!

1. According to the passage, what caused the flourishing of the postal systems in Han Dynasty?
2. As the development of mobile phones, fewer and fewer people choose to write letters. What's your opinion on continuing or giving up the letters written by hand?

92 Swimsuit（泳衣）

The first recorded use of a form of swimsuit dates back to Greece in 350 BC. The early 1800s witnessed a revolution with regards to the swimsuit when Americans would travel in groups to the beach for recreation. The first swimsuits consisted of bloomers and black stockings. By 1855, drawers were worn to avoid exposure when wearing the swimsuit. Improvements were made regarding the cut of the suit. By the 1880s, the "Princess" cut was introduced, consisting of a blouse and trousers in one piece. The skirts were replaced by cotton-like trousers. A separate skirt fell below the knee and buttoned at the waist to conceal the figure. A ruffled cap or a straw hat was also worn to complete the attire.

1. According to the passage, what caused the revolution to the swimsuit in the early 1800s?
2. The swimsuit has developed a lot. What's your opinion on whether it will just stop or continue to have new forms? Why?

55

93 Socks（袜子）

Studies suggest that the first socks were made during the Stone Age using animal skins, which people tied around their ankles. By the eighth century BC, Greek poet Hesiod wrote about "piloi", which are socks made from matted animal hair. The Romans began wrapping their feet in strips of leather or woven fabric. By the second century AD, they were wearing "udones", which were sewn from woven fabric and pulled over one's foot. The first real knit socks were discovered in the Egyptian tombs between the third and sixth centuries. In Europe, socks were basically strips of cloth or hide that were wrapped around the legs and feet. They were called "leggings". The year 1938 saw the development of nylon, which led to the blending of two or more fabrics, a process that is currently used in the production of socks. Other fabrics such as acrylic①, polyester②, polyamide③ and spandex④ are also used.

1. According to the passage, why did people call socks "leggings" in Europe?
2. Nowadays, socks are in all shapes and sizes. Do you think it's necessary? Why?

The trend to produce colourful socks led to a major blend of new styles, patterns and looks. Coloured socks are often a part of school uniforms and are worn by sports teams on the field.

Socks are available in all shapes and sizes. There are knee-high socks, toe socks, short socks, anklets, over-the-knee socks, bare socks and more.

① acrylic 丙烯酸纤维　② polyester 聚酯纤维；涤纶　③ polyamide 聚酰胺　④ spandex 氨纶（弹性纤维）

1 AD – 1600 AD
（公元1年-公元1600年）

94　Paper（纸）

Paper was first invented in China. Cai Lun, an official at the imperial court of the Han Dynasty during the early second century improved the technique. Paper is a thin material that is produced by pressing together moist fibres such as cellulose pulp, which is obtained from wood, rags or grasses, and then dried to create sheets.

Paper is used in several ways. While its most common use is for writing and printing, it is also widely used as a packaging material, in many cleaning products, in a number of industrial and construction processes and even as a food ingredient, mainly in Asian cultures. The word paper is derived from "papyrus", which is the ancient Greek name of the "Cyprus papyrus" plant. Papyrus is a thick, paper-like material that is produced from the pith of the Cyprus papyrus plant, which was used in ancient Egypt and other Mediterranean cultures for writing, much before paper was invented.

The knowledge of papermaking moved from China to Japan, then to Korea in 610 AD, where it was commonly made from mulberry bark and Gampi (Japanese shrubs). Later, it was made from bamboo and rice straw.

> 1. According to the passage, what was used for writing in ancient Egypt and other Mediterranean cultures before paper was invented?
> 2. Nowadays we call for paperless office. Do you think paper will disappear in the future? Give at least one reason.

95 Abacus（算盘）

Abacus was developed by the Chinese and was used for calculations like addition, subtraction, multiplication and division as well as fractions and square roots.

A Chinese abacus has a wooden frame that is divided into two parts. These are separated by a beam with an upper deck of two rows of beads and a lower deck of five rows of beads. A series of vertical rods allows the wooden beads to slide freely. Traditionally, the abacus was made of wood or stone.

1. According to the passage, can we use an abacus to know the answer to $\sqrt{(37+76)\times 36/2}$?
2. Have you learned how to use an abacus? If not, will you be interested in learning it? Give your reasons.

96 Toothbrush（牙刷）

The modern toothbrush was invented in 1938, much later than the toothpaste. Early forms of the toothbrush have existed since 3000 BC. Ancient civilisations used a thin twig with a frayed end to clean their teeth.

The bristle toothbrush, similar to the type used today, was not invented until 1498 in China. The stiff, coarse hair from the back of a pig's neck were used as bristles, attached to handles made of bone or bamboo.

Nylon bristles were introduced in 1938 by Dupont de Nemours. The first nylon toothbrush was called "Doctor West's Miracle Toothbrush".

1. According to the passage, which one is later invented, the toothbrush or the toothpaste?
2. Why was the first nylon toothbrush called "Doctor West's Miracle Toothbrush"?

97 Vault（拱顶）

A vault with regards to architecture is a structural aspect that consists of an arrangement of arches, usually forming a ceiling or roof. In ancient Egypt, brick vaulting was used for drains.

Vaults have various forms. The simplest form is the tunnel vault, also known as the barrel vault, which can be described as a "continuous arch". The weight of such a vault requires thick, supportive walls with limited gaps. As the height of a tunnel vault should increase along with its width, there is a practical limit to its size.

1. What is necessary to make a barrel vault?
2. Do you think vaults have a long history? Why?

98 Hydrometer（液体比重计）

A Greek scholar, Hypatia of Alexandria, has been given the credit for inventing the hydrometer during the late fourth century or early fifth century.

A hydrometer is an instrument that is used to measure the specific gravity or density of liquids. It is usually made of glass. It consists of a cylindrical stem and a bulb weighted with mercury or lead to make it float upright. The liquid to be tested is poured into a tall container, often a cylinder, and the hydrometer is gently lowered into the liquid until it floats freely. The point at which the surface of the liquid touches the stem of the hydrometer is noted. Hydrometers usually contain a scale inside the stem, so that the specific gravity can be read directly.

1. According to the passage, what makes a hydrometer float upright?
2. How can we know about the gravity or density of liquids by using a hydrometer?

99 Map（地图）

The first great attempt to make mapping realistic came in the second century AD with Claudius Ptolemy. He was an astronomer and astrologer obsessed with making accurate horoscopes, which required precisely placing someone's birth town on a world map.

Ptolemy gathered documents detailing the locations of towns, and he augmented that information with the tales of travellers. He devised a system of lines of latitude and longitude, and plotted some 10,000 locations—from Britain to Europe, Asia and North Africa. Ptolemy even invented ways to flatten the planet (like most Greeks and Romans, he knew the Earth was round) onto a two-dimensional map.

The Babylonian map of the world, a clay tablet created around 700 to 500 BC in Mesopotamia, a resemblance of a map, depicted a circular Babylon at the centre, bisected by the Euphrates River and surrounded by the ocean. And it wasn't really for navigation. Centuries later, the Romans drew an extensive map of their empire on a long scroll, which was not realistic.

As the Renaissance dawned, maps began to improve. Christopher Columbus' voyage to America was partly due to Ptolemy. Columbus carried a map influenced by the ancient Roman's work, even though he discovered the map had a few problems.

1. How did Ptolemy get the information for the map?
2. What do you think of Ptolemy's devising a system of latitude and longitude?

1 AD – 1600 AD
（公元1年－公元1600年）

100 Paper Money（纸币）

The first recorded use of paper money was believed to be in China during the seventh century. The Chinese used paper currency to avoid carrying heavy and cumbersome metallic coins for transactions. Similar to making a deposit at a modern bank, individuals would transfer their coins to a trustworthy party and then receive a note denoting how much money they had deposited. The note could then be redeemed for currency at a later date. They continued this process for more than 500 years before the practice spread to Europe in the 17th century.

1. Why did the ancient Chinese people use paper currency during the seventh century?
2. Do you like using paper money or mobile payment systems? Why?

101 Cannon（大炮）

The cannon was first invented in China and is among the earliest forms of gunpowder artillery. In the Middle East, it is debated that the hand cannon was first used during the Battle of Ain Jalut between the Mamluks and Mongols in 1260. The first cannon in Europe was probably used in Iberia during the 11th and 12th centuries.

The word cannon is derived from the old Italian word "cannone", which means a "large tube". It has a truncated cone with an internal, cylindrical bore for holding an explosive charge and a projectile.

The world's earliest known cannon dates back to 1282 when China was in Yuan Dynasty, under the rule of Mongols.

1. Why do people call cannons "large tubes"?
2. Is the invention of the cannon a good idea, or a bad technology?

102 Pretzel（椒盐卷饼）

In 610, while baking bread, an Italian monk decided to create a treat to motivate his distracted students. He rolled out ropes of dough and twisted them so that they looked like hands crossed on the chest in prayer, and he baked them. The monk called his snacks "pretiola", which is Latin for "little reward". When the pretiola arrived in Germany, it was called "brezel". Because of its religious roots, the pretzel has long been considered a good-luck symbol. German children wear pretzels around their necks on New Year's Day. The pretzel serves as an emblem for bakers since the 12th century.

Pretzels now have different shapes. Salt is the most common seasoning for pretzels, complementing the washing soda or lye treatment that gives pretzels their traditional "skin" and flavour through the Maillard reaction. Other seasonings include sugars, chocolate, glazes, seeds and/or nuts.

1. What does the word "seasoning" in the second paragraph mean in Chinese?
2. List a kind of food that is regarded to bring good luck to people in China?

1 AD – 1600 AD
（公元1年—公元1600年）

103 Gun（枪）

The first recorded gun firing occurred in China during the 13th century. The first firearm was the fire lance, the prototype of a gun. The fire lance was invented in China during the 10th century. The term "gun" may refer to any sort of projectile weapon ranging from large cannons to small firearms, including those that are handheld. A gun is a normally tubular weapon or any other device designed to discharge projectiles or other materials.

1. When and where did the first recorded gun firing occur?
2. There are several types of guns around the world. Please find them out and list some.

104 Gunpowder（火药）

Ancient Chinese alchemists were trying to find a formula for immortality and ended up creating gunpowder. It was a mixture of sulphur①, saltpetre (potassium nitrate)② and charcoal. When the Chinese found out that it exploded, they began to use it for fireworks. After a few hundred years, people started using it for war. They first used it at the beginning of a war to try and scare the enemies. They realised that if gunpowder was exploded near people, they might die, so they started using it in wars as an explosive.

1. What was gunpowder in ancient China according to the passage?
2. What can gunpowder be used to do?

In the 12th and 13th centuries, gunpowder spread to the Arab countries, then Greece, other European countries and finally all over the world.

① sulphur 硫磺　② saltpetre (potassium nitrate) 硝酸钾

105 Velvet（法兰绒）

Velvet is a woven fabric in which the cut threads are evenly distributed. It has a short, dense pile that gives it a unique feel. Velvet can be made from synthetic or natural fibres. This fabric originated in Kashmir. It is associated with nobility. Velvet is woven on a special loom that simultaneously weaves two thicknesses of the material. The weaving technique dates back to as early as 2000 BC in Egypt. Because of its fairly intricate manufacturing process, velvet was an extremely expensive fabric and continues to be fairly expensive even today.

1. Why was velvet very expensive?
2. How to understand "Velvet is associated with nobility"?

106 Rocket（火箭）

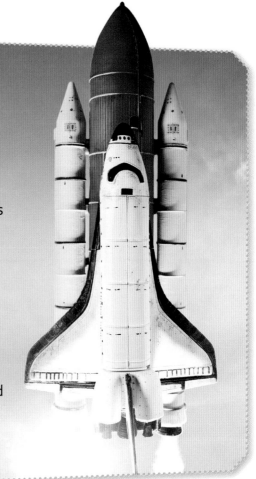

The first rocket was invented around 1100 AD in China. These rockets were mainly used as weapons and fireworks. It was during the 1920s that rocket societies emerged. By the 1930s and 1940s, professional rocket engineering began. Three pioneers began working independently on developing rockets to reach space. Konstantin Tsiolkovsky and Hermann Oberth were the first to come up with many essential principles and realise that the rocket could be used as a means to travel into space. This created a curiosity about rocketry and space travel. In 1926, Robert Goddard launched the first liquid-propellant rocket. Due to its secrecy, the 1926 rocket did not influence later developments.

1. What were rockets mainly used as before in China?
2. Name one fairy tale that is related to space travel in ancient times in China.

107 Dress（裙子）

It is not known who invented the dress or when they invented it. It cannot be credited to a single inventor. Chinese legend has it that the Emperpor Huangdi set rules for people's dressing, including the dress.

Women often wore tunics overlaid with fabrics in ancient European societies, right up to the Middle Ages. Back then, wearing fabric became a way for nobility to showcase their status in society. By adding more fabric to tunics, females could show off their position in society. This led to the evolution of dresses. The tunic was gradually reduced to being peasant-wear and gowns became the symbol of high society. Clothing became so important during the Middle Ages that laws were passed about the types of clothing that people of various social strata could and could not wear.

1. How could people know about a woman's status in ancient societies?
2. Do you agree "the development of clothing can reflect the development of a country"? And how do you understand this sentence?

Spinning Wheel（纺车）

The spinning wheel has been in use since ancient times when spinning was done on a spindle. The spindle was a stick with a stone or some weight attached to it. In China, the earliest evidence of a spinning wheel can be traced back to Han Dynasty which was mentioned in *Fangyan*. According to Irfan Habib, the spinning wheel was introduced in India from Iran in the 13th century. In France, the spinning wheel was not used until the mid-18th century.

To use a spinning wheel, cleaned wool or cotton is first carded. This means that cotton is spread on one card and combed with another, until the fibres are all facing one direction. The carding is done with hand cards and coarse nail brushes that are about 12 inches long and five inches wide. The cotton is then twisted loosely and finally spun into yarn. It is then taken off in fleecy rolls that are about 12 inches long and three quarters of an inch thick. These short cardings are twisted on the spinning wheel into a loose thread, about the size of a candlewick. These threads are wound on reels or bobbins and finally spun into the finished yarn. Several types of fibres can be spun on a simple spinning wheel.

1. Where did the spinning wheel appear earliest in Chinese history?
2. Before using a spinning wheel, what need to do first?

109 Rifle (来福枪)

Rifled firearms date back to the 15th century Europe. In 1610, artist, gunsmith and inventor Marin le Bourgeoys developed the first "flintlock" for King Louis XIII of France. A rifle is a firearm that is designed to be fired from the shoulder. It has a barrel with a helical groove or a pattern of grooves cut into the barrel walls. The trigger releases a spring-loaded mechanism that causes a flint to strike a steel surface. The spark ignites gunpowder and propels a spherical bullet.

Formerly, rifles only fired a single projectile with each squeeze of the trigger. Modern rifles are capable of firing more than one round per trigger squeeze. Some are fully automatic and others are limited to fixed bursts of two, three or more rounds per squeeze.

1. What causes a flint to strike a steel surface?
2. From the passage, we know some rifles are fully automatic and others are not. What kind of rifles do you think are better? Why?

110 Lace (蕾丝)

Lace is a delicate, openwork fabric made of yarn or thread, often found on fancy attire. It is patterned with open holes in the work, which can be made by machine or hand. The holes can be formed by removing threads or pieces of cloth from a previously woven fabric, but more often, these open spaces are created as a part of the lace fabric.

Lacemaking is an ancient craft that can be traced all the way back to the early 16th century. While many have debated over the inventor of lace, it is hard to give the credit to one person as lacemaking evolved from various other techniques. Lace gained a lot of popularity during the 1500s, when it started to make an appearance in both fashion and home decor.

1. How long has lacemaking been as an ancient craft?
2. Do you think lacemaking is important? Why?

Open woven fabrics and fine nets that had a lace-like effect are known to have existed for centuries. Originally, lace was created using linen, silk, gold or silver threads. Today, lace is often made with cotton thread.

111 Spectacles（眼镜）

Evidence suggests that glasses first appeared in Pisa, Italy, around 1268. They were formed from two simple convex-shaped glass/crystal stones. Each of these was surrounded by a frame and given a handle. These were connected together through the ends of their handles by a rivet. They were not an invention, but an idea based on the simple glass stone magnifier. Someone took two, existing, mounted stones and connected them with a rivet. The first pair of glasses was invented by a lay person who wanted to keep the process a secret in order to make a profit. Spectacles are used to correct your vision if you cannot clearly see things that are at a distance or even nearby.

Though the eyeglasses existed for a while, it was troublesome to keep them on the eye. Finally, in 1730, an optician named Edward Scarlett found a solution to keeping them on by using rigid side pieces that could be hooked behind the ears.

1. What are spectacles used to do?
2. What is the passage mainly about?

Today, glasses are available in many types, based on their primary function, but they also appear in combinations such as prescription sunglasses or safety glasses that enhance magnification.

1 AD – 1600 AD
（公元1年—公元1600年）

112 Screwdriver（螺丝刀）

The inventor of the screw is unknown, although screw-shaped tools were a common item since the first century. The first screwdrivers were used to unscrew corks on wine and olive oil bottles. Initially, they were made of wood, but now they are made of metal for extra strength, durability and stability. To accompany the screwdriver, metal screws and nuts were created to fasten two objects together in the 15th century. The screwdriver was designed to insert and tighten bolts or screws, or loosen and remove bolts or screws.

1. What were the first screwdrivers used to do?
2. What would happen if there were no screwdrivers in our daily life?

113 Electricity（电）

Electricity was never invented because it is a form of energy that occurs naturally. Rubbing amber on a cat's fur attracted light objects. This was known to ancient cultures around the Mediterranean. Around 600 BC, Thales of Miletus became the earliest researcher of electricity. He rubbed fur with other objects and found them to attract each other. What he actually discovered was "static electricity".

In 1600, English physician William Gilbert studied the relationship between electricity and magnetism in detail. He was able to find the lodestone effect due to static electricity, which was produced by rubbing amber. Gilbert named it after the Latin word "electricus", meaning "of amber". He is said to be the father of modern electricity. In 1646, Thomas Browne's *Pseudodoxia Epidemica* included the English words "electric" and "electricity".

1. How can "static electricity" be created?
2. Static electricity happens easily in life and sometimes it can be annoying. What can you do to avoid it?

114 Pencil（铅笔）

Pencils were first produced in the 16th century. A large deposit of graphite was discovered in Borrowdale, England, during the first half of the 1500s. Local residents cut the graphite into sticks and used them to mark their sheep. The graphite was misidentified as lead, a word that has been connected with pencils ever since, even though modern pencils do not contain lead. Graphite left a darker mark than lead, but it was soft and brittle, so it had to be held. Therefore, it was inserted into hollow, wooden sticks.

Later, an Italian couple, Simonio and Lyndiana Bernacotti, invented the pencil in its modern form around 1560. They hollowed out a stick of juniper wood and placed a graphite stick inside. Another technique was developed where a graphite stick was inserted into two wooden halves that were glued together. This basic technique remained in use even 400 years later.

1. What was graphite mistaken for?
2. Why was graphite put into hollow, wooden sticks?

115 Watch（表）

The first watch was invented in the 1500s in Germany by a locksmith named Peter Hanlein. It was like a portable clock. However, it was so heavy that it had to be held by a belt, which was worn around the waist.

A watch is a mechanical device, which is powered by winding a main spring. This spring turns the gears that are responsible for moving the hands. These hands keep account of time with a rotating balance wheel.

The history of the watch spans 500 years, of which, most of the time was devoted to refining the mechanical watch.

1. Why was the first watch worn around the waist?
2. As the popularization of mobile phones, do you think the watch will be replaced one day?

116 Printing Press (印刷术)

A printing press is a device for evenly printing ink on paper or cloth. Typically used for texts, the invention and spread of the printing press is widely regarded as one of the most influential events in human history. The early printing press was operated by hand. The earliest documented evidence of printing dates back to the second century, when the ancient Chinese started using wooden blocks to transfer images of flowers on silk.

During the reign of Qingli, 1041–1048, Bi Sheng, made movable type by cutting stick clay into very thin characters and baking them. He placed the types onto a previously prepared iron plate covered with a mixture of pine resin, wax, and paper ashes. He then placed it near the fire to warm it. When the paste melted slightly, he took a smooth board and pressed it over the surface, so that the block of type became as even.

In 1440, German inventor Johannes Gutenberg invented a printing press process that, with refinements and increased mechanisation, remained the principal means of printing until the late 20th century. This allowed the first mass production of printed books. During the 19th century, other inventors created steam-powered printing presses that did not require a hand operator. Today's printing presses are electronic and automated, and can print faster than before.

The printing press is one of the most important inventions in history. It has ensured that books, newspapers, magazines and other reading materials are produced in great numbers, and it plays an important role in promoting literacy among people.

1. Why did Bi Sheng take a smooth board and press it over the surface of the paste?
2. Why is the printing press one of the most important inventions in history?

117 Corset (紧身衣)

A corset is a close-fitting, stiff piece of clothing that provides shape to a woman's torso. It was a popular garment during the 16th century in Europe. However, the term "corset" was used from the 19th century. Prior to that, the corset was called "bodies", "a stiff bodice" or "a pair of stays".

The first corset was invented during 1500–1550. It was made from stiff materials like whalebone, horn and buckram, and was referred to as "whalebone bodies". A stay is placed vertically in the centre of the torso to keep it straight.

French queen Catherine de Medici, wife of King Henry II, introduced the corset to France.

1. What was the corset called before 19th century?
2. Do you have any comments on this invention?

118 Teapot (茶壶)

The teapot was invented in the 1500s by clay potters in Yixing County, Jiangsu province of China during the Ming Dynasty period. Initially, teapots were small, unglazed brown or red pieces of pottery with wide bases, spouts and handles, and were used exclusively for brewing tea. These pots were able to withstand extreme heat when hot, boiling liquids were poured into them. In the 1600s, teapots were brought by Dutch importers from China to Europe along with chests of tea leaves. After a century of experimentation, European potters finally managed to produce a quality teapot similar to the heat-resistant Chinese ones.

Since then, the teapot has evolved from plain clay to fine glazed porcelain and to translucent and exquisite bone china. In spite of these new varieties, authentic Yixing teapots continue to be highly coveted by tea enthusiasts.

1. What was the teapot initially used for?
2. Do you like the European porcelains or the authentic Yixing teapots? Give your reasons.

119 - Microscope (显微镜)

During the 1590s, two Dutch spectacle makers, Zacharias Jansen and his father Hans, started experimenting with glass lenses. They put several lenses in a tube that led to a very important discovery. The object near the end of the tube appeared to be greatly enlarged, much larger than any simple magnifying glass could achieve by itself.

Their first microscopes were more of a novelty and not very useful, since the maximum magnification was only around 9x (times) and the images were blurry to a certain extent. The early Jansen microscopes were compound and used a minimum of two lenses. The objective lens was positioned close to the object. It produced an image that was picked up and magnified further by the second lens, which was called the "eyepiece".

Antony van Leeuwenhoek was the first man to make and use the modern microscope. Leeuwenhoek ground and polished a small glass ball into a lens with a magnification of 270x, and used this lens to make the microscope.

Because it had only one lens, his microscope is now commonly referred to as a single-lens microscope. Its convex glass lens was attached to a metal holder and focused with the help of screws. Leeuwenhoek constructed a total of 400 microscopes during his prolific lifetime.

1. What does the word "it" in line 4 paragraph 2, refer to?
2. In which field can we use the microscope? Why?

1 AD – 1600 AD
(公元1年–公元1600年)

120 Heeled Shoes（高跟鞋）

High-heeled shoes were first worn by men! They were used in the 16th century by Persian soldiers who rode on horseback. The shoes offered stability in the stirrups to the soldiers so they could use their bow and arrows more efficiently. Later, in 17th century Europe, they caught on as a fashion statement for the aristocracy. Around the 1630s, women started adopting masculine fashion trends and began wearing high heels.

The year 1533 saw the first women's heel that was designed to lengthen the legs. The invention of high heels as a fashion statement could be accredited to the rather petite Catherine de Medici. Heels were adopted by the European aristocracy during the 1600s as a sign of status.

1. What were heels adopted by the Europeans as during the 1600s?
2. Do you think it is useful for Persian soldiers to wear high-heel shoes? Why?

121 Stockings（长筒袜）

Stockings are also known as hosiery, hose and popularly as "nylons". They are used as a covering for legs and feet. Early references to stockings date back to the ancient Greeks. Workmen and slaves wore hosiery in ancient times, and Roman women wore a short sock called a "soccus" within their homes. Silk or cotton stockings were also worn in Japan and China for centuries.

The soccus evolved into stockings in 12th century Europe. After 1545, knitted stockings came into fashion. Interestingly, several pairs of silk stockings were sometimes worn during winter, even though knitted stockings offered more warmth!

1. When and where did the soccus evolve into stockings?
2. Which do you prefer to wear, silk stockings or knitted stockings? Why?

1 AD – 1600 AD
（公元1年—公元1600年）

122 Bullet（子弹）

A bullet is a projectile (often pointed) metal cylinder that is shot from a firearm. After 1249, it was understood that gunpowder could be used to fire projectiles out of the open end of a tube. The earliest firearms were large cannons, but small and personal firearms appeared only in the mid-14th century. Early projectiles were stone or metal objects that could fit down the barrel of the firearm. Lead and lead alloys came to be used by 1550.

Bullet is usually a part of an ammunition cartridge, the object that contains the bullet. It is inserted into the firearm. Bullets are made from a variety of materials. Lead or a lead alloy is the traditional bullet core material. There are many other materials that are used in bullets today, including aluminium, bismuth[①], bronze, copper, plastics, rubber, steel, tin and tungsten[②].

1. What were early projectiles?
2. What do you think of the bullet? Give your reasons.

123 Hat（帽子）

Hats have been around for a long time. They were used for protection and as a fashionable accessory. Additionally, they are known to have a long history as markers of status, occupation and even political affiliation.

One of the first hats to be depicted was found in a tomb painting at "Thebes". It showed a man wearing a labourer-style straw hat. In the late 17th century, women's headgear rose to fame. Women's hats were starkly different from the ones that men wore around that time.

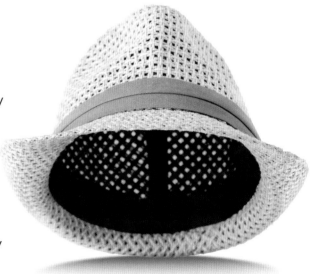

A maker of women's hats was called a "milliner". This term dates back to 1529. It referred to the products that made Milan and northern Italian regions famous.

1. What were hats used for according to the passage?
2. Do you like wearing a hat? Why?

---

① bismuth 铋 ② tungsten 钨

124 Clothes Iron（熨铁）

Various objects have been used for centuries to remove wrinkles and press clothing. However, there was a time when this could only be afforded by rich people. Slaves or servants were hired to do the work. Around 400 BC, the Greeks used a "goffering" iron to create pleats on linen robes. The goffering iron was a round bar that was heated before use.

By the 10th century, Vikings from Scandinavia had early irons made of glass. They used a linen smoother to iron pleats. This smoother was warmed by being held near steam and then rubbed across the fabric. The iron first appeared in Europe in the 1300s. It constituted a flat piece of iron with a metal handle and was held over a fire till it turned hot.

1. How was a linen smoother used?
2. Do you think it is useful to use clothes iron? Why?

125 Gregorian Calendar（格里高利历）

The calendar that we refer to today is the Gregorian calendar. It was proposed by Aloysius Lilius, a physician from Naples and was adopted by Pope Gregory XIII to correct the errors in the older Julian calendar. This calendar was officially declared by Pope Gregory XIII in 1582.

In the Gregorian calendar, the tropical year is approximated as 36597/400 days = 365.2425 days. Therefore, it takes approximately 3,300 years for the tropical year to shift one day with respect to the Gregorian calendar. The approximation 36597/400 is achieved by having 97 leap years in every 400 years.

1. According to the passage, how did the Gregorian calendar evolve?
2. What kind of calendars do we use in China? Which kind do you prefer?

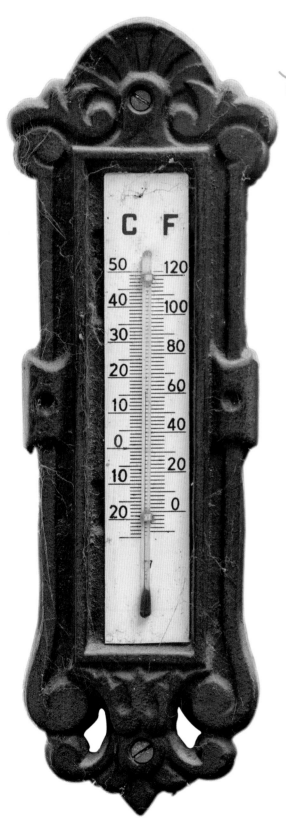

126 Thermometer（温度计）

Thermometers are used to measure temperature. This is carried out using materials that change in a particular manner when heated or cooled. In a mercury or alcohol thermometer, the liquid expands when it is heated and contracts when it is cooled. Before the thermometer, an instrument called the thermoscope was used. It was a thermometer without a scale. A thermoscope only showed the differences in temperatures. For example, it could show that something was getting hotter.

Several inventors simultaneously invented various versions of the thermoscope. In 1592, Galileo Galilei invented a simple water thermoscope, which enabled different temperatures to be measured. Today, Galileo's invention is called the "Galileo Thermometer".

In 1612, an Italian inventor Santorio became the first inventor to put a numerical scale on his thermoscope. It was the first, basic, clinical thermometer, as it was designed to be placed in a patient's mouth for checking the temperature. In 1654, the first, enclosed thermometer was invented by Duke Ferdinand II of Tuscany. He used alcohol in it. However, it was still inaccurate and did not use a standardised scale. Daniel Gabriel Fahrenheit invented the first mercury thermometer with a standardised scale in 1714.

1. What is the difference between the thermometer and the thermoscope?
2. What are these various versions of the thermoscope in common?

1601 AD – 1800 AD
（公元1601年-公元1800年）

127 Railroad（铁路）

"Wagonways" were the first railroads that came into existence. They were used in Germany in 1550. These railed roads comprised wooden rails over which horse drawn wagons or carts moved with greater ease as compared to dirt roads. Wagonways paved the start of modern railroads.

By 1776, the wooden rails and wheels on the carts were replaced by iron. Wagonways evolved into tramways and spread throughout Europe. Even then, horses were used to pull these carts. In 1789, Englishman William Jessup created the first wagon with flanged wheels. The flange was a groove that allowed the wheels to have a better grip on the rail. This was a beneficial design that aided the later versions of locomotives.

1. Why was the flange a beneficial design?
2. How did roads and wagons evolve according to the passage?

1601 AD – 1800 AD
(公元1601年—公元1800年)

128 Cork（软木塞）

One does not know exactly when humans thought about using the bark of a tree to plug a bottle. The material used to make the cork is found on the bark of the cork oak tree. It is made from dead cells that accumulate on the outer surface. Cork bottle stoppers were found in Egyptian tombs that were thousands of years old. Ancient Greeks used corks to make fishing net floats, sandals and bottle stoppers. Romans used corks for various purposes, including life jackets for fishermen. For centuries, Mediterranean cottages have been built with cork roofs and floors to prevent them from getting too hot or cold.

A cork tree can be harvested when it is around 20 years old. The use of cork as a stopper grew wildly popular, leading to an increase in the cultivation of cork trees.

1. What did ancient Greeks use corks to do?
2. Why did the use of cork lead to an increase in the cultivation of cork trees?

129 Bow Tie（领结）

Bow ties came to be worn since the 1700s. Earlier, Europeans wore scarves around their necks to hold the tops of their shirts in place at the collar. The black bow tie being a part of the "black tie" attire was first used in 1886, when the tuxedo was invented by Pierre Lorillard V. Earlier, the fashion was to wear tailcoats with white bow ties. The tuxedo is one such attire that has survived all eras and the only accurate complement to such a suit is the bow tie.

1. Why did Europeans wear scarves around their necks?
2. If a man wants to attend a formal dinner party with his boss, what may he wear together with his tuxedo? Why?

130 Submarine (潜水艇)

Sketches of a submarine were first created by Leonardo da Vinci. In 1578, William Bourne, a British mathematician, drew plans for a submarine. However, the first submarine was created in 1620 by Cornelius van Drebbel, a Dutch inventor. He tightly wrapped a wooden rowboat in waterproof leather and added air tubes with floats to the surface to provide oxygen. His submarine had no engines, so the oars went through the hull at leather gaskets.

The first submarine to be used for military purposes was built in 1775 by American inventor David Bushnell. His submarine was a one-man, wooden submarine that was powered by hand-turned propellers.

Two American inventors, John P. Holland and Simon Lake, developed the first true submarines in the 1890s. The US Navy purchased submarines built by Holland, while Russia and Japan chose Lake's designs. Their submarines used petrol or steam engines for surface cruising and electric motors underwater. The first nuclear-powered submarine, the "USS Nautilus", was launched in 1954.

1. Who invented the first submarine?
2. How was the first submarine developed?

131 Steam Engine (蒸汽机)

1. What was the first steam pump called?
2. How was the steam engine invented?

Thomas Savery was the first to invent a steam pump in 1698. He called it "water by fire". In 1712, Thomas Newcomen invented an effective and practical steam engine. It consisted of a piston or cylinder that moved a huge piece of wood to drive the water pump. This engine was used for more than 50 years. James Watt improved (1763–1775) Newcomen's engine with a seperate condenser and further provided a rotary motion for driving machinery, which enabled factories to be sited away from rivers, and accelerated the pace of Industrial Revolution.

The invention of the steam engine was a difficult process. In September 1825, the Stockton & Darlington Railroad Company was the first railroad to carry both goods and passengers on regular schedules using locomotives designed by English inventor George Stephenson. It pulled six loaded coal cars and 21 passenger cars with 450 passengers over 14 km in about one hour.

George Stephenson is considered to be the inventor of the first steam locomotive engine.

1601 AD – 1800 AD
（公元1601年-公元1800年）

132 Telescope（望远镜）

Contrary to popular belief, Galileo did not invent the telescope. Three others have also claimed to invent the telescope — Hans Lippershey, Zacharias Jansen and Jacob Metius. However, Lippershey was the first to apply for a patent. The telescope that Lippershey invented could magnify up to three times only. The instrument consisted of a positive lens at one end of a narrow tube and a negative lens at the other end.

Galileo was the first to use the telescope for the study of astronomy in 1609. He could see mountains and craters on the Moon as well as the Milky Way. His observations helped him discover that the Sun had sunspots and Jupiter had its own moons.

1. What did Galileo do for the study of astronomy?
2. Have you ever used a telescope? What can you do with a telescope?

133 Tie（领带）

A necktie or tie is a long piece of cloth that is worn around the neck or shoulders. It rests under the shirt collar and is knotted in the front of the neck. A tie is worn for style.

The actual year of invention of the tie is debatable. It is believed that the tie was first used by Croatian soldiers during the Thirty Years' War that started in the 1600s. The word "cravate" was used to describe the handkerchief that Croatian soldiers tied around their necks during the war. This made it easy to identify them. Silk ties were strictly reserved for officers, while the soldiers wore ties of ordinary materials.

King Louis XVI of France began sporting a lace cravate during 1646, when he was just seven years old. This set the trend of neckties among the French nobility of that era. It was only during the period of 1910 to 1919 that neckties began to resemble the modern ones we see today.

1. Why do people wear a tie?
2. Why did Croatian soldiers and officers wear ties?

134 Barometer（气压计）

The word barometer is derived from the Greek words "baros", which means weight, and "metron", which means measure. The barometer is used to measure air pressure. Barometer uses the principle of a vacuum to measure the weight of air.

The first working barometer was created in 1643 by Evangelista Torricelli. He worked with and studied the writings of Galileo, just before Galileo's untimely death in 1642. Torricelli used those findings to construct the first barometer, which made use of water to measure the air pressure during that time.

1. What do "baros" and "metron" mean?
2. How did Torricelli construct the first barometer?

135 Blood Transfusion（输血）

Blood transfusion is the process of receiving blood products directly into your blood stream through your veins. Transfusions are used for various medically-associated conditions to replace lost components of blood. Early transfusions used whole blood, but modern medical practice only uses components of blood, such as red blood cells, white blood cells, plasma, clotting factors and platelets.

The first research on blood transfusion dates back to the 17th century when British physician William Harvey described the circulation and properties of blood in 1628. The first blood transfusions were also tried during this period, but they often failed and proved dangerous to humans.

The first successful blood transfusion was performed by a British physician Richard Lower in 1665. He let a dog almost bleed to death and revived it by transfusing blood from another dog through a tied artery. In 1667, Jean-Baptiste Denis, King Louis XIV's physician, performed blood transfusion from an animal to a human. He transfused blood from a sheep to a 15-year old boy and later to a labourer. Both these transfusions were successful.

1. Compared to modern medical practice, what did early transfusion use?
2. From the passage, we know that blood transfusion didn't succeed until 1665. What can we learn from this fact?

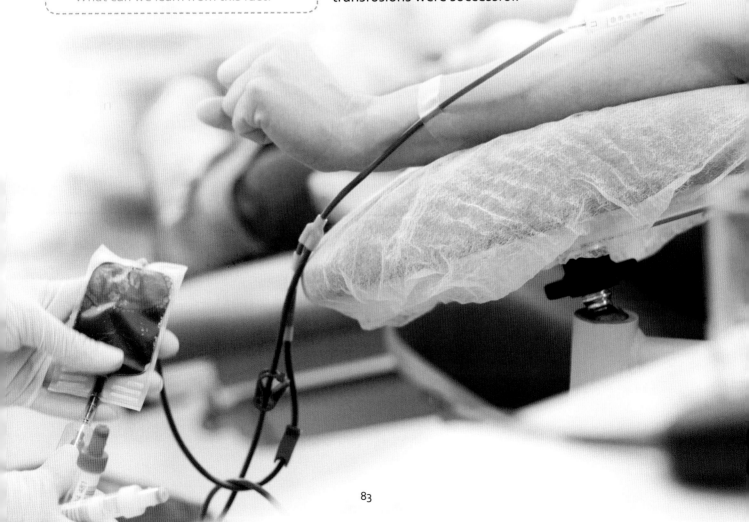

136 Parachute（降落伞）

The first time someone thought of an idea similar to a parachute was in 1514, when Leonardo da Vinci sketched its design in his notebook. Many years later, another man by the name of Fausto Veranzio published his own design, which was strikingly similar to that of Da Vinci. Veranzio went on to explain exactly how this device worked by jumping from a high place, because he believed that it would work. However, the first man to successfully try the parachute is said to be a Frenchman named J.P. Blanchard. He dropped a little dog sitting in a basket all by itself from a hot air balloon in flight and watched it safely land on the ground. He even claimed that he used the parachute himself in 1793, but broke his leg when he touched the ground.

1. Who first thought of the idea of a parachute?
2. What was the possible reason for thinking of a parachute at that early time?

The first man to use the parachute regularly and on recorded documentation was another Frenchman named André Jacques Garnerin. The first time that he went parachuting was in 1797, when he jumped off a height of 600 m but landed safely and securely.

137 Refrigerator（冰箱）

Oliver Evans first invented the refrigerator in 1805. However, William Cullen invented the process in 1748 and Jacob Perkins added improvements to the refrigerator in 1834. The first refrigerator was introduced in 1834. By 1880, there were over 3,000 patents for refrigerators.

Albert Einstein also patented an invention of the refrigerator. In 1903, he invented an eco-friendly refrigerator, with no moving parts, that did not use electricity.

Earlier, people used iceboxes to keep their food cool and prevent it from getting spoilt. Iceboxes were lined with metal and insulated with straw, sawdust or cork. Blocks of ice were put on top so that cold air would circulate downwards to keep the food cool. A tap would be used to drain the melting water.

> 1. How did people keep food cool in earlier times?
> 2. Do you like the invention of the refrigerator? Why or why not?

138 Water Frame（水力纺纱机）

Richard Arkwright was a barber and wig maker in Bolton, England around 1750. He learnt that he would become rich if he could invent a machine that would spin cotton fibre into yarn quickly and easily. He teamed up with a clockmaker named John Kay. By the late 1760s, they had a workable machine that spun four strands of cotton yarn at the same time. Arkwright patented this machine in 1769 to stop others from copying his invention.

The water frame spins 96 strands of yarn together. It was similar to the machines installed in the mills in Derbyshire and Lancashire that were powered by water wheels. As a result, they were called "Water Frames". Currently, it is the only machine of its kind in the world that is complete. Arkwright's machines did not require skilled operators.

> 1. How many strands of yarn can the water frame spin together?
> 2. What do you think of water frame? Please find one supporting detail in the passage.

139 Razor（剃须刀）

During prehistoric times, clam shells or flint were used for removing hair. Later, between 3000 BC and 6000 BC, razors evolved from clam shells and flint to a more sophisticated device. Archaeologists discovered circular razors made of bronze in ancient Egyptian burial chambers.

The first safety razor was conceptualised around 1770 by Frenchman Jean-Jacques Perret. It consisted of a sharpened, straight razor with a wooden guard. During this period, shaving was done by professional barbers.

During 1895, a salesman named King C. Gillette came up with an idea for a double-edged safety razor that would be made of cheap, disposable blades, which would not have to be sharpened. He was inspired by his colleague, William Painter, who suggested that one way to make money was to produce something cheap that people would need to buy repeatedly. However, it was difficult to mass produce such blades. In 1901, Gillette was helped by a MIT graduate William Nickerson and by 1903, they had successfully invented a thin, sharp blade which could be produced in large quantities.

1. What did people find in Egyptian burial chambers?
2. What improvements do you want to make on modern razors?

140 Mayonnaise（蛋黄酱）

Some culinary historians observed that a mayonnaise-like mixture of olive oil and egg was frequently consumed by ancient Egyptians and Romans. The mayonnaise that we have today is an emulsion of oil, egg, lemon juice and/or vinegar along with different types of seasonings. This was developed by a chef from France.

Mayonnaise was invented by Duke de Richelieu's chef in 1756. When the Duke conquered a Mediterranean island, the chef wanted to prepare a victory sauce but lacked the essential ingredient – cream. So, he created a new sauce using eggs and olive oil, and named it "mahonnaise".

The first ready mayonnaise was sold at Richard Hellman's New York deli in 1905. In 1912, it was marketed and called "Hellman's Blue Ribbon Mayonnaise".

1. What is today's mayonnaise?
2. Why do people who want to lose weight have to say no to mayonnaise?

141 Accelerometer (加速度计)

An accelerometer is used to measure different kinds of acceleration, i.e., the rate at which velocity changes. It was initially used to validate the principles of Newtonian physics.

The first accelerometer was invented by an English physicist George Atwood in 1783. It measured linear acceleration: for example, the rate at which an object falls. A spring system is used to measure the accelerating force, which provides acceleration using Newton's famous second law, i.e., force equals mass times acceleration. Later, accelerometers were designed to measure circular or twisting acceleration, like that of a weight attached at the end of a string. Here, the acceleration depends on the radius of the circle of the spinning object.

1. An accelerometer is used to measure different kinds of acceleration, isn't it?
2. If the mass of A is 3 kg, the acceleration is 1 m/s², what is the force (N)?

142 Carbonated Water (气泡水)

Carbonated water, also known as club soda, soda water, sparkling water, seltzer water or fizzy water is water into which carbon dioxide gas has been dissolved under pressure. The first drinkable man-made glass of carbonated water was invented by Joseph Priestley in 1767.

In the late 18th century, J. J. Schweppe developed a process to manufacture carbonated mineral water using the same process discovered by Joseph Priestley, thus founding the Schweppes company in Geneva in 1783. In 1799, Augustine Thwaites founded Thwaites' Soda Water in Dublin. Today, carbonated water is made by passing pressurised carbon dioxide through flavoured or regular water.

1. Why did Joseph Priestley found the Schweppes Company in 1783?
2. What may be the possible reasons for the popularity of the soda water?

143 Spinning Jenny（珍妮纺纱机）

James Hargreaves invented a device to spin cotton in 1764. He named his invention the "Spinning Jenny" after his wife. This invention was important as it was the first improvement on the spinning wheel. It paved the way for the Industrial Revolution, which led people to leave the countryside and move to major cities. Therefore, it may be said that Hargreaves influenced European and eventually American lifestyles. His invention led to the setting up of textile mills all across Europe and even the USA.

Originally, the spinning jenny used eight spindles instead of the single one that was used on the spinning wheel. A single wheel on the spinning jenny controlled eight spindles, which created a weave using eight threads that were spun from a corresponding set of rovings. Later models had up to 120 spindles.

1. What did Hargreaves do on 12 July, 1770?
2. Why is the spinning jenny important in the history? Give at least two reasons.

On 12 July, 1770, Hargreaves patented a 16-spindle spinning jenny. His invention decreased the need for labour. The only drawback was that his machine produced thread that was too thick to be used for warp threads and could only be used for weft threads.

Sandwich(三明治)

Hillel the Elder was said to be the first person that ate sandwich during the first century BC. This kind of sandwich consists of chopped nuts, apples, spices, bitter herbs and wine, which are sandwiched between two matzos. Since he was the first known person to come up with the idea and considering his influence and status, the Hillel Sandwich was named after him.

Food historians attribute the conception of the sandwich to John Montagu, the fourth Earl of Sandwich. Montagu was an avid gambler. It is said that on 3 November, 1762, during a 24-hour gambling streak, he instructed a cook to prepare food in a manner that would not require him to stop playing his ongoing game. The cook offered him sliced meat that was placed between two pieces of toast. This meal did not need any utensils and could be eaten using only one hand, which left the other hand free to continue the game. While the Earl got credited for the invention, the name of his cook remains unknown.

1. According to the passage, when was the first sandwich eaten?
2. Why are sandwiches so popular in the modern society?

145 Smallpox Vaccine（天花疫苗）

In 1796, scientist Edward Jenner inoculated a healthy, 8-year-old boy with cowpox. It is a disease that is caused by a virus closely resembling variola. Cowpox usually affects small mammals such as wood mice, but the virus can spread to other animals as well, especially cattle.

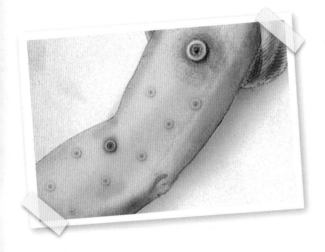

Jenner's experiment was successful. His patient did not contract smallpox, even when he was deliberately exposed to variola. By 1800, smallpox vaccinations were common, mainly because they caused fewer side effects and deaths than variolation with smallpox itself.

In 1989, the World Health Assembly declared smallpox eliminated.

1. According to the passage, why did the author say Jenner's experiment was successful?
2. Do you want to be a scientist like Edward Jenner in the future? Why or why not?

146 Steam Boat（蒸汽船）

Marquis Joffroy D'abans, a Frenchman, was the first to attempt to build a steam boat. He launched his steam boat in 1776 and called it "Palmipède". He placed a small, wood-fired steam engine on a river boat. The engine's rocker arms were connected to paddles that looked like duck feet when flopping in water. He attempted to sail it along the River Seine in Paris, but the boat did not move.

1. Pyroscaphe wasn't the first steam boat which sailed successfully, was it?
2. What's the difference between Palmipède and Pyroscaphe?

The first steam-powered ship, "Pyroscaphe", was a paddle steamer powered by a Newcomen steam engine. It was built in France in 1783 by Marquis Claude de Joffroy and his colleagues as an improvement of the 1776 Palmipède. Its first demonstration was on 15 July, 1783, where the Pyroscaphe travelled on the River Saône for 15 minutes before the engine failed.

In 1807, Robert Fulton, an American engineer and inventor, developed the first commercially successful steam boat with passengers travelling along the Hudson River.

147 Iron Bridge
（铁桥）

Abraham Darby cast the world's first iron bridge in Coalbrookdale. He was a local ironmaster and his bridge was erected across the River Severn in 1779. Shrewsbury architect, Thomas Pritchard first suggested to ironmaster John Wilkinson in 1773 that an iron bridge should be built over the Severn. Pritchard designed the bridge. The construction was completed in 1779. The world's first iron bridge was opened on New Year's Day in 1781. It cost over £6,000.

The bridge is built from five cast iron ribs spanning across 30.6 metres or 100 feet. 378 tonnes of iron were used to construct the bridge. Almost 1,700 individual components were used for the construction, the heaviest weighing 5.5 tonnes. Individual components were cast to fit each other instead of following standard sizes. There were discrepancies of up to several centimetres between those components that were identical in different locations. The opening of the bridge resulted in several changes in lifestyle for those living in the settlement. The condition of the roads around the bridge improved in the years after its construction.

1. Each of the individual component weighed 5.5 tonnes, didn't it?
2. According to the passage, people could travel very comfortably on the roads around the bridge right after its construction, couldn't they? Why?

148 Hot Air Balloon（热气球）

On 4 June, 1783, the Montgolfier brothers built a balloon made of silk and lined with paper that was 10 metres in diameter. They launched this giant balloon from the marketplace in Annonay, France. During their first attempt, nobody was aboard on this balloon. It rose to a magnificent height of 1,600-2,000 metres and stayed aloft for 10 minutes. It travelled for a distance of about 2 km.

The Montgolfier brothers' next step was to put a person in the basket and fly it once again. On 15 October, 1783, they launched a balloon on a tether with Jean-François Pilâtre de Rozier, a chemistry and physics teacher, aboard. He stayed up in the air for almost four minutes.

About a month later, on 21st November, Pilâtre de Rozier and Marquis d'Arlandes, a French military officer, made the first free ascent in a hot air balloon. They flew from the middle of Paris all the way to the suburbs. They covered what was then considered an impressive distance of about 9 km in 25 minutes.

The next major point in balloon history was on 9 January, 1793. Jean Pierre Blanchard became the first man to fly a hot air balloon in North America, with George Washington's presence at the launch.

1. How long did the hot air balloon stay aloft during the Montgolfier brothers' second attempt?
2. According to the passage, can you guess why a chemistry and physics teacher would like to be the first person to fly in the hot air balloon?

1601 AD – 1800 AD
（公元1601年—公元1800年）

149 Thresher（脱粒机）

Thresher was first invented by Scottish mechanical engineer Andrew Meikle around 1784 to separate grain from stalks and husks. For thousands of years, grain was separated by hand with flails, which was very difficult and time-consuming.

1. What are the advantages of threshers?
2. Can you list some grain products that are processed by threshers before coming to your dining tables?

Early threshing machines were hand-fed and powered by horses. They were small, about the size of an upright piano. The later versions were steam-powered, driven by a portable or traction engine.

150 Cotton Gin（轧棉机）

Eli Whitney invented a simple machine that influenced the history of the USA in 1793. He invented a cotton gin that was particularly popular in southern states. The South became the cotton producing part of the country because Whitney's cotton gin successfully pulled out seeds from cotton bolls.

1. According to the passage, why did the South become the cotton producing part of the country?
2. Why did the author draw the conclusion that the cotton gins influenced the history of the USA in 1793?

The cotton gin was a simple invention. Cotton bolls were put into to the top of the machine. Then, all one had to do was to turn the handle, which turned the cotton through the wire teeth that combed out the seeds. Finally, cotton was pulled away from the wire teeth and out of the cotton gin.

151 Ball Bearing（滚珠轴承）

Philip Vaughan, in 1749, invented the ball bearings to be used in carriage wheels. And now, the entire industrial world runs on them. Two circular grooved rings known as "races", one fitting inside the other, provide the contact between the two components moving relative to one another. Metal balls fitted between the races keep the friction of the movement as low as possible.

It is known that a circle gives the least amount of continuous contact with its surroundings and it is this principle on which the ball bearings work. If the weight being carried by the bearings becomes too heavy, the balls are then replaced with rollers; these have lines of contact with the races rather than just points and, therefore, possess a larger load-carrying capacity.

Keeping the balls in alignment and keeping dirt out is crucial for their smooth operation. Friction can generate enormous amounts of heat at the speed at which many bearings operate. The traditional all-steel ball bearing is being replaced with high-performance one made of incredibly resilient silicon nitride① ceramic balls fitted into races made from fabric-reinforced phenolic resin, giving substantial improvements in performance.

1. On which principle can the ball bearings work?
2. Why should the metal balls fitted between the races keep the friction of the movement as low as possible?

---

① silicon nitride 氮化硅

152 Primary Cell Battery（原电池）

A battery produces electricity from a chemical reaction. It is used to power various appliances.

The initial method of generating electricity was by creating a static charge. Otto von Guericke constructed the first electrical machine in 1663. It consisted of a large sulphur globe, which attracted feathers and small pieces of paper when it was rubbed and turned. Guericke proved that the sparks generated were truly electrical.

The first suggested use of static electricity was the so-called "electric pistol", which was invented by Alessandro Volta. An electrical wire was placed in a jar filled with methane gas. When an electrical spark was sent through the wire, the jar exploded. In 1831, Michael Faraday showed how a copper disc provided a constant flow of electricity when it revolved in a strong magnetic field.

In 1836, John F. Daniell, an English chemist, further researched the electro-chemical battery and created an improved cell that produced a steadier current than Volta's device. Until then, all batteries were composed of primary cells, which meant that they could not be recharged.

1. Who is the inventor of the first suggested use of static electricity?
2. Do you think that the "electric pistol" is dangerous? Why?

1801 AD – 1850 AD
（公元1801年-公元1850年）

153 Protractor（量角器）

> 1. Why did US naval captain Joseph Huddart invent a more multifaceted protractor?
> 2. With the fast development of technology, do you think it is still necessary for students to learn how to use protractors? Why?

Protractors are mathematical drawing instruments that are used to draw and measure angles. They have existed since ancient times, dating back to the 13th century. During that period, European instrument makers created an astronomical observing device called the "torquetum①" that also consisted of a semicircular protractor.

A more multifaceted form of the protractor was later designed for plotting the position of a ship on navigational charts. This was invented by US naval captain Joseph Huddart in 1801. This instrument was called a three-arm protractor or station pointer, and was composed of a circular scale that connected to three arms. The centre arm was fixed, whereas the outer two were rotatable, which made them capable of being set at any angle relative to the centre one.

Another similar instrument used by marine navigators is the course protractor. It works as an effective tool that measures the angular distance between magnetic north and the course plotted on a navigational chart.

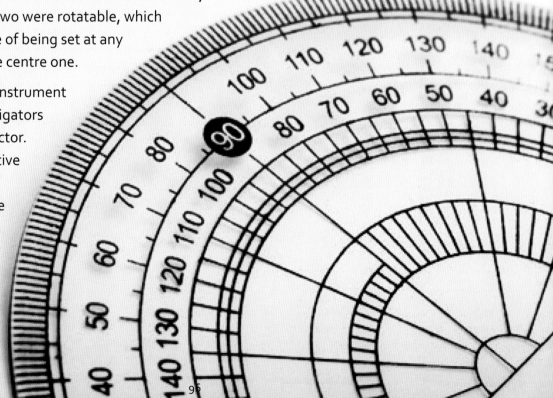

① torquetum 赤基黄道仪

154 Hang Glider（滑翔机）

A hang glider is devised to carry a human passenger who is suspended beneath its sail in the air. Hang gliders are usually launched from a high point. They slowly drift down to a lower point.

The modern history of hand gliding begins with the English inventor Sir George Cayley. By 1799, Cayley established a basic design for gliders. In 1804, he flew his first successful model glider. In 1853, Cayley achieved the first successful human glider flight. The next important pioneer was German inventor Otto Lilienthal. In the 1890s, he built around 18 gliders, which he flew himself.

1. According to the history of the hang glider, who can be considered to be essential?
2. According to the passage, guess the reason why hang gliders can be used as a kind of entertainment.

155 Quinine（奎宁）

Quinine is a substance that is found in the bark of the cinchona tree. It was used as a cure for malaria. However, this was later replaced by modern medicine.

In 1817, two French scientists, Pierre-Joseph Pelletier and Joseph Bienaimé Caventou, extracted quinine from the bark of the South American cinchona tree. Before they discovered that quinine cured malaria, the bark of the cinchona tree was first dried, ground to a fine powder and mixed into a liquid before being consumed. This medicine stopped people from shivering in lower temperatures. In 1820, quinine was extracted from the bark of the tree and used as a replacement for the ground bark.

1. What did the two French scientists do before quinine was used as the cure for malaria?
2. Do you know any other medicine that is extracted from plants?

156 Tin Can (锡罐)

It is said that the tin can was invented by the Frenchman Philippe de Girard. He passed the idea to a British merchant named Peter Durand, who revolutionised food preservation when he patented the tin can in 1810. In 1813, John Hall and Bryan Dorkin opened the first commercial canning factory in England. In 1846, Henry Evans invented a machine that manufactured tin cans at a rate of 60 cans per hour. That was a significant increase over the previous rate of only six cans per hour.

Henry Evans invented a method for making a can from a single motion. After a year, Allen Taylor patented his machine-stamped method of producing tin cans.

> 1. List all the stages that the tin can was invented and manufactured. (Including people and time.)
> 2. How long did it take the tin can from being invented to being manufactured with a patent?

157 Solar Cell (太阳能电池)

A solar cell is any device that converts light energy into electrical energy through the process of "photovoltaics". In 1839, French physicist Antoine-César Becquerel developed the technology of solar cells. While experimenting with a solid electrode in an electrolyte solution, he saw a voltage develop when light fell on the electrode. This helped him understand the photovoltaic effect.

> 1. How long did it take from the conception of the solar cell to the commercially available invention?
> 2. What do you think gave Antoine-César Becquerel an idea to understand the photovoltaic effect?

The earliest solar cells and panels were not extremely efficient. In 1941, Russell Ohl created the silicon solar cell. In 1954, three American researchers, Gerald Pearson, Calvin Fuller and Daryl Chapin, designed a silicon solar cell capable of six per cent energy conversion under direct sunlight. In 1956, the first solar cells were commercially available.

158 Tractor（拖拉机）

The first traction engines were developed during the 1850s. These were widely used for agricultural activities. Prior to these, portable engines powered by steam were used to drive mechanical farm machinery using a flexible belt. The first tractors that were used for ploughing were run by steam engines. Two engines were fixed on opposite sides on a farm to haul a plough back and forth in the space between them using wire cables. Under favourable soil conditions, steam engines were used to haul ploughs directly. Steam-powered engines were used until the 20th century.

The word tractor is derived from the Latin word "trahere", which means to pull. The word "tractor", meaning "an engine meant for wagons or other implements", was coined in the year 1901.

In 1892, in Northeast Iowa, John Froelich invented the first successful gasoline-powered engine that could be driven backwards and forwards. During that period, steam-powered engines were used for threshing wheat. John Froelich knew how to operate such equipment, which helped him invent the gasoline-powered engine.

1. Which appeared earlier, gasoline-powered engines or steam-powered engines?
2. What helped John Froelich invented the gasoline-powered engines?

159 Reaper（收割机）

The reaper was a horse-drawn farm implement that was invented by Cyrus Hall McCormick in 1831. It was created to cut small grain crops. The mechanical reaper replaced the manual cutting of crops, which was done using scythes and sickles. This machine was used to cut down wheat quickly and efficiently.

The reaper harvested wheat and other grains, and could cut around 15 acres a day. Prior to this invention, only three acres could be cut manually. The harvester gained immediate popularity and farmers found the reaper extremely helpful, because it reduced labour costs and the danger of weather destroying crops. The reaper cut the stalks, which fell on a platform. A worker pushed them on the ground with a rake. This task required eight to ten workers.

In 1847, McCormick built a reaper factory in Chicago, next to the Chicago River. In 1848, the McCormick factory manufactured 700 machines. By 1850, this number had doubled. By 1868, 10,000 reapers were made every year. These machines were also exported to other countries. With each progressing year, the reapers got heavier, stronger and better. The Chicago factory was considered as one of the greatest industrial establishments in the USA.

160 Stethoscope（听诊器）

Prior to the invention of stethoscope, a physician would listen to a patient's heart by placing his ear over the patient's chest. In 1816, a physician named René Laennec was called to examine a young woman who was believed to have had a heart disease. Based on the medical procedure of the time, Laennec tapped his hand on the patient's back and tried to listen to the resulting sound. However, as the patient was plump, he could not hear anything.

To avoid putting his ear on the young woman's chest, Laennec came up with a simple solution, whereby he rolled a piece of paper into a cylinder and used that to listen to the patient's heartbeat. Laennec

1. What are the advantages of a reaper compared to the manual cutting of crops?
2. Why was the Chicago factory considered as one of the greatest industrial establishments in the USA?

later created a new instrument from a hollow, wooden cylinder that he called "stethoscope". This word comes from the Greek words "stethos" meaning chest and "skopos" meaning examination.

1. Why did Laennec come up with the idea of rolling a piece of paper to listen to the patient's heartbeat?
2. Do you think the invention of stethoscope was great? What convenience did it bring?

161 Camera（照相机）

During the fifth century BC, a Chinese philosopher named Mo Di observed that a pinhole can form an inverted and focused image when light passes through it and into a dark area. Early cameras were plain boxes that focused light through a pinhole. During the 15th century, good-quality glass lenses were used to focus these images. By the 19th century, chemicals such as silver nitrate① allowed a permanent image to be preserved, which established the modern science of photography.

Over time, cameras were developed, rather than being invented by a specific individual. In 1685, Johann Zahn described, but could not build a device that would capture images. In 1825, Nicephore Niepce used bitumen to create the first actual photograph.

The first practical, portable camera was built by Louis Daguerre in 1837. The first practical camera that could be used by a layman was invented by George Eastman in 1888.

1. What was the Chinese philosopher Mo Di's contribution to the invention of a camera?
2. What does the sentence "cameras were developed, rather than being invented by a specific individual" mean in the passage?

① silver nitrate 硝酸银

162 Bicycle（自行车）

The earliest bicycle was a wooden, scooter-like device called a "celerifere". It was invented around 1790 by Comte Mede de Sivrac of France. In 1816, Baron Karl von Drais de Sauerbrun of Germany invented a model with a steering bar attached to the front wheel. He called it a "draisienne". It consisted of two same-sized wheels and the rider sat between the two wheels. The draisienne did not have any pedals. In order to move ahead, one had to propel the bicycle forward using their feet. He exhibited his bicycle in Paris on 6 April, 1818.

Kirkpatrick MacMillan, a blacksmith from Dumfriesshire, Scotland, invented the first bicycle with foot pedals between the 1830s and 1840s. However, he never patented it and his idea did not become popular.

French carriage makers Pierre and Ernest Michaux invented a bicycle in the 1860s, which was an improvement to the previous one. Many early bicycles called "velocipedes", meaning "fast foot", had huge front wheels. It was believed that the bigger the wheel, the faster you could go. Early tyres were wooden and were soon replaced by metal tyres; solid rubber tyres were added much later. A chain with sprockets was added to the bicycle during the 1880s. This was called the "safety bicycle". Air-filled tyres were also added during the 1880s. The gear system that we currently see was added in the 1970s.

1. How did a draisienne move ahead?
2. Can you draw the time line of the invention of the bicycle?

163 Suspenders（吊裤带）

Suspenders are worn to hold up trousers. The first suspenders date back to 18th century France. Back then, they were strips of ribbons that were attached to the buttonholes of trousers.

During the 1820s, British clothing designer Albert Thurston began to mass manufacture "braces", which is the British word for suspenders. Instead of metal clasps that clasped to the trousers's waistband, these "braces" used leather loops that were attached to buttons on the pants. At the time, British men wore very high-waist trousers and did not use belts.

> 1. According to the passage, what's the difference between braces and suspenders?
> 2. Do you like wearing suspenders? Why?

164 Matchstick（火柴）

A matchstick is a wooden stick that contains a coat of phosphorus[①] mixture at one end. This mixture is ignited by striking the matchstick against the rough surface of the matchbox to produce a flame.

Matchsticks originated around 3500 BC. The Egyptians developed a small, pinewood stick that was coated with a combustible mixture of sulphur. In 1827, the "friction" matchsticks that we use even today were invented by an English chemist named John Walker. The match head had a mixture of antimony sulphide[②], potassium chlorate[③], starch and gum.

In 1844, "safety" matches were invented by a Swedish man called Gustaf Erik Pasch and were improved by John Edvard Lundstrom a decade later. These matchsticks ignited only when they were stroked at a specific place.

> 1. How can we ignite the matchstick?
> 2. Why don't we use matchsticks very often nowadays?

① phosphorus 磷　② antimony sulphide 硫化锑　③ potassium chlorate 氯酸钾

165 Macintosh Raincoat（麦金托什雨衣）

In 1823, Scottish chemist Charles Macintosh patented a method for making waterproof garments by using rubber dissolved in coal-tar① naphtha② for cementing two pieces of cloth together. The now famous Macintosh raincoat is named after Charles Macintosh. This raincoat was first made using the method developed by Charles Macintosh. It led to the creation of the first waterproof fabric. However, the fabric had flaws and was easy to puncture when it was seamed. The natural oil in wool caused the rubber cement to deteriorate. When it was cold, the fabric turned stiff and when it was hot, it turned sticky.

With the invention of vulcanised rubber in 1839, Macintosh's fabric improved as it could withstand temperature changes.

1. What was the method for making waterproof garments?
2. After 1839, would the fabric turn stiff when it was cold? Why?

166 Bus（公交车）

Blaise Pascal, a mathematician, inventor, physicist, philosopher, author and general savant, came up with the idea of a bus and secured finance from his friends in the nobility. The system began with seven horse-drawn vehicles running along regular routes. In 1824, John Greenwood established the first modern omnibus service. As the keeper of a toll gate in Pendleton on the Manchester-to-Liverpool turnpike, he purchased a horse and cart with several seats, and began an omnibus service between those two locations. His service did not require prior booking; the driver would pick up or set down passengers anywhere upon their request.

1. According to the passage, did the omnibus have regular stops? Why?
2. If you were one of the passengers, what would you say to John Greenwood after taking the bus?

① coal-tar 煤焦油   ② naphtha 石脑油

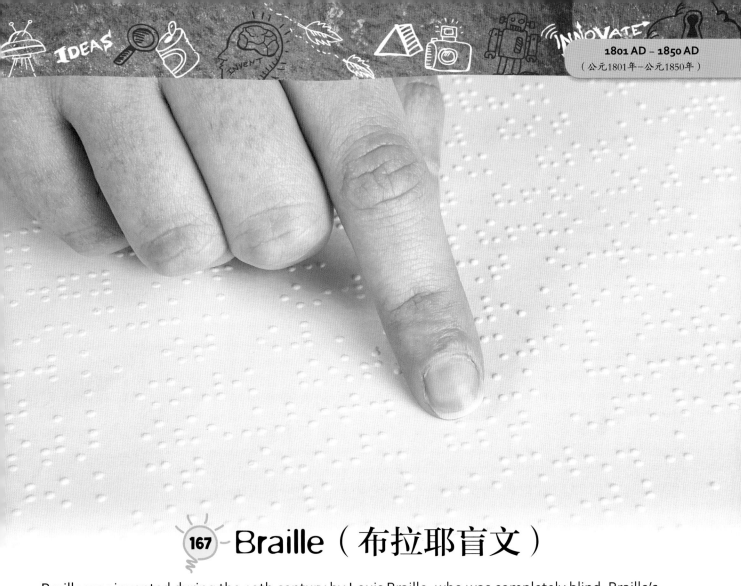

167 Braille（布拉耶盲文）

Braille was invented during the 19th century by Louis Braille, who was completely blind. Braille's story goes back to when he was three years old. He injured his eye while he was playing. Though he was offered the best medical attention available at the time, it was not sufficient. An infection soon developed and spread to his other eye, because of which he turned blind in both eyes. It is because of this accident that we have Braille today.

During that time, a system of reading was already in place for the blind; it consisted of tracing a finger along raised letters. However, this system was extremely slow and it was difficult to understand simply by touching the relatively complex letters of the alphabet. Therefore, many people struggled to master the embossed letter system.

Braille is read by passing one's fingertips over characters made from an arrangement of one to six embossed points. The positions of these points represent different alphanumeric characters. Braille can be written with a "Braillewriter" that is similar to a typewriter. A pointed stylus can be used to punch dots through paper using an instrument called a "Braille slate", which has rows of small cells as a guide. Braille has been adapted to almost every known language and is an essential tool for those who are visually challenged.

1. What happened to Louis Braille when he was three years old?
2. Where can you see Braille in our daily life? What do you think of it?

1801 AD – 1850 AD
（公元1801年—公元1850年）

168 Sewing Machine(缝纫机)

In 1790, Thomas Saint, a British inventor, was the first to patent a design for the sewing machine. His machine was to be used only on leather and canvas. However, a working model was not built. In 1814, Austrian tailor Josef Madersperger presented his first sewing machine. The development of this machine had begun in 1807.

In 1830, French tailor Barthélemy Thimonnier had patented a sewing machine that sewed straight seams using a chain stitch. By 1841, Thimonnier had a factory consisting of 80 machines that sewed uniforms for the French Army.

The first sewing machine had all the disparate elements of the previous half-century of innovation. The modern sewing machine was a device built by an English inventor, John Fisher, in 1844 for processing lace materials. His machine was quite similar to the devices that were built by Isaac Merritt Singer and Elias Howe in the following years. However, due to the botched filing of Fisher's patent at the patent office, he did not receive due recognition for his contribution towards the modern sewing machine.

1. According to the passage, did John Fisher receive the recognition? Why?
2. If you were one of the judges, would you give him the recognition? Why?

169 Handbag（手提包）

?
1. What did professional luggage makers do?
2. According to the passage, what do you think helped develop the bags?

Purses and handbags have their origins in the early pouches that were used to carry seeds, religious items and medicine. With the onset of the railroad, bags were experiencing a revolution.

In 1843, there were over 3,000 km of railway lines in Great Britain, because of which many people began travelling by train and women also became more mobile. Professional luggage makers turned the skills of horse travel into those for train travel. Soon, the term "handbag" was coined to describe the new, handheld luggage bags. The famous Hermès bags were founded in 1837 by Thierry Hermès, a harness and saddle maker, while Louis Vuitton was a luggage packer for the Parisian rich.

170 Harvester（联合收割机）

The combine harvester is considered to be the modern harvester of wheat. It is called a combine because it "combines" the job of the header and thresher, which were its predecessors. The combine originated in the Midwest, USA, but also had a significant impact on the Northwest and specifically in the state of Washington.

The first combine was made by Hiram Moore in 1836 and was extremely advanced. It was successful because it made farming safer, more profitable and helped food reach many people. However, through the 1800s, the header and the thresher were used individually.

?
1. Why do we call the modern harvester of wheat a combine harvester?
2. If you were the farmer in 1836, would you buy a combine harvester? Why?

171 Morse Code（摩斯密码）

In 1836, Samuel Morse showed how a telegraph system can transmit information over wires. The information was passed as a series of electrical signals. Short signals were referred to as "dits", which were represented as dots. Long signals were referred to as "dahs", which were represented as dashes. With the invention of radio communications, an international version of Morse code came to be used on a large scale.

Morse code depends on specific intervals of time between dits and dahs, between letters and words. The rate of transmitting the Morse code is measured in words per minute. The word "Paris" is used as the standard length of a word. To transmit the word requires 50 units of time. If you transmitted the word five times, you would be transmitting at 5 WPM. An experienced Morse code operator can transmit and receive information at 20–30 WPM.

In 1844, Morse sent his first telegraph message from Washington DC to Baltimore, Maryland, USA; by 1866, a telegraph line had been laid across the Atlantic Ocean from the USA to Europe. Although the telegraph had fallen out of widespread use by the start of the 21st century, and was replaced by the telephone, fax machine and Internet, it had laid the groundwork for the communication revolution that led to all those innovations.

1. Why did the telegraph fall out of widespread use by the start of the 21st century?
2. Do you want to learn Morse code? Why?

1801 AD – 1850 AD
(公元1801年—公元1850年)

172 Telegraph（电报）

In 1794, the non-electric telegraph was invented by Claude Chappe. This system was visual and it used semaphore, a flag-based alphabet. It depended on a line of sight for communication. With time, the optical telegraph was replaced by the electric telegraph. In 1809, a crude telegraph was invented in Bavaria by Samuel Soemmering. He used 35 wires with gold electrodes in water and at the receiving end of 2,000 feet, the message was read by the amount of gas caused by electrolysis. In 1828, the first telegraph in the USA was invented by Harrison Dyar who sent electrical sparks through chemically treated paper tape to burn dots and dashes.

In 1832 Samuel Morse, assisted by Alfred Vail, came up with the idea for an electromechanical telegraph, which he called the "recording telegraph". This commercial application of electricity was made real by their construction of a crude working model in 1835–1836. This instrument was probably never used outside Morse's rooms, where it was operated in many demonstrations.

The telegraph was further refined by Morse, Vail and a colleague, Leonard Gale, into a working mechanical form in 1837. The flow of electricity through the wire was interrupted for shorter or longer periods by holding down the key of the device. These dots or dashes were recorded on a printer or could be interpreted orally.

In 1838, Morse perfected his sending and receiving code, as well as founded a corporation, making Vail and Gale his partners.

1. According to the passage, how did the optical telegraph work?
2. Name at least three kinds of telegraphs that are mentioned in this passage.

173 Stamps（邮票）

In 1680, English merchant William Dockwra and his partner Robert Murray established the London Penny Post, a mail system that delivered letters and small parcels within London for one penny. The postage was prepaid and this payment was confirmed by the use of a hand-stamp that marked the mailed package. The first stamp was black and it was named "Penny Black". It was released on 6 May 1840. Later, several other countries used this idea. To avoid confusion, they put the name of their country on the stamp.

Most other countries have different portraits as stamps, but Great Britain is the only country to use just a picture of the Queen!

The Penny Black had the left profile of Queen Victoria's head, which remained on all British stamps for the following 60 years. Rowland Hill is credited to create the first stamp.

1. What was the London Penny Post?
2. According to the passage, what are the stamps used to do?

174 Suspension Bridge（悬索桥）

Around 150 years ago, William Hamilton Merritt was the first person to visualise a bridge over the Niagara River. He planned and built the first Welland Canal, making it possible for ships to avoid the Niagara Falls.

In 1846, the governments of Upper Canada and the state of New York, USA consented and started the formation of two companies with the ability to construct a bridge at or near the falls. The companies were called the Niagara Falls Suspension Bridge Company of Canada and the International Bridge Company of New York. Both companies jointly built and owned the bridge. In the autumn of 1847, the bridge companies commissioned Charles Ellet Jr. to construct a bridge at a location selected by the companies along the Niagara River.

The suspension bridge provided an inexpensive solution to the problem of long spans over navigable streams or at other sites where it is otherwise difficult to build piers. British, French, American and other engineers of the late 18th and early 19th centuries faced serious problems of stability and strength against wind forces and heavy loads, failures resulting from storms, heavy snows and droves of cattle.

1. Who was the first person to visualise a bridge over the Niagara River? Did he build it eventually?
2. According to the passage, what's the advantage and disadvantage of the suspension bridge?

1801 AD – 1850 AD
（公元1801年—公元1850年）

175 Stapler（订书机）

1. What caused people to design modern paper fastening devices?
2. What kind of stapler mentioned in this passage do you think is similar to the stapler that we are using today?

The first stapler dates back to 18th century France. The first handmade stapling machines or "fasteners" were developed for King Louis XIV of France during the 1700s. The increase in the use of paper during the 19th century created the need for an efficient paper fastener. Modern paper-fastening devices began with the patent of the first paper fastener on 30th September, 1841, by Samuel Slocum. This basic device stuck pins on paper to hold them together. In 1879, a machine that inserted and clinched a single, pre-formed, metal staple entered the market. It was called "McGill's Patent Single Stroke Staple Press".

176 Aluminium（铝）

Aluminium salts were first said to be used by Ancient Greeks and Romans as astringents for dressing their wounds and for fixing dyes. In 1761, Guyton de Morveau suggested that the base alum should be called "alumine". Scientists suspected that an unknown metal existed in alum as early as 1787, but they did not have any method to extract it until 1825. In 1809, Humphry Davy acknowledged the existence of a metal base in alum, which he eventually called aluminium. Hans Christian Oersted, a Danish chemist, was the first to produce tiny amounts of aluminium. After two years, Friedrich Wöhler, a German chemist, developed another manner to obtain the metal. By 1845, he produced sufficient samples to determine some of aluminium's basic properties. Wöhler's method was improved upon in 1854 by Henri Étienne Sainte-Claire Deville, a French chemist. Deville's process enabled aluminium to be produced commercially.

1. When were the scientists able to extract aluminum?
2. Do you know what people use "aluminium" to do today? Please give at least one example.

177 Voltmeter(电压表)

A voltmeter is defined as an instrument that is used to measure the difference in electrical potential between two points of an electric circuit. The first voltmeter was called a "galvanometer". In 1824, a French physicist and mathematician, Andre-Marie Ampere, invented the first prototype galvanometer. It was the first instrument that could measure the electrical current of a conductor. His invention had its root in the concept of the first galvanometer, as reported by Johann Schweigger on 16 September, 1820. The mechanism of this invention was based on the principle that a magnetic needle would turn away from its position if an electric current was present nearby. The earliest model of the galvanometer or voltmeter consisted of a compass, which was surrounded by a wire coil.

The early galvanometers were not very accurate or consistent. Modern voltmeters measure the potential difference between two points on an electric circuit. The term "galvanometer" was coined after the surname of an Italian researcher of electricity, Luigi Galvani. In 1791, Galvani discovered that an electric current could cause a frog's leg to jerk.

1. What is a voltmeter used for?
2. Have you ever used a voltmeter? If so, when and where did you use it?

1801 AD – 1850 AD
（公元1801年—公元1850年）

178 Dirigible（飞艇）

A dirigible or airship is an aircraft that consists of a cigar-shaped gas bag, which is filled with a lighter-than-air gas to provide lift, a propulsion system, a steering mechanism and a gondola that can accommodate passengers, crew and cargo. French inventor Henri Giffard built a steam-powered airship in 1852. However, with the invention of the gasoline engine in 1896, airships were practical to use.

Alberto Santos-Dumont from Brazil was the first to construct and fly a gasoline-powered airship in 1898. During 1910, the "Deutschland" became the first commercial dirigible. Between 1910 and 1914, German dirigibles flew over 1,72,000 km and carried 34,028 passengers and crew without any harm.

1. What kind of gas is filled in the dirigible? Why do people fill the gas bag with it?
2. Why aren't the dirigible being widely used nowadays?

179 Safety Pin（别针）

The safety pin was invented by a man named Walter Hunt in New York in 1849. It was made from a small piece of metal, which was a combination of copper, iron, aluminium, gold, silver and platinum. These metals were heated and moulded into a small piece.

One afternoon, Walter Hunt had to think of a way to pay a 15-dollar debt. He was sitting at his desk and twisting a piece of wire while trying to think. He twisted the wire for three hours and realised that he had created something different and useful. He called it the safety pin. However, he did not invent the safety pin; he only improved it. It was not the first pin, but it was the first to have a clasp that prevented the sharp edge from poking someone. The first pin was invented by the ancient Greeks, Italians and Sicilians.

1. What is special about Walter Hunt's safety pin?
2. Did Walter Hunt invent the safety pin on purpose or not? How can we know this?

180 Fax (传真)

Alexander Bain invented the first kind of technology to send an image through a wire. While working on an experimental fax machine between 1843 and 1846, he was able to synchronise the movement of two pendulums through a clock. With that, he managed to scan a message on a line-by-line basis. Frederick Bakewell improved on Bain's invention and created an image telegraph that is very similar to today's fax machine.

Bakewell replaced Bain's pendulums with matching rotating cylinders, which allowed a clearer image through better synchronisation. This telegraph would lift the image from the cylinder with a stylus and place it on the other cylinder through a similar stylus onto chemically-treated paper. Bakewell's image telegraph was successful. It was the first crucial step towards a commercially practical method to send images over a wire.

1. Why do we say that Bakewell's invention was the first crucial step towards a commercially practical method to send images over a wire?
2. According to this passage, what factors should be considered in choosing a fax machine?

Following Bain's achievements, a group of inventors made several revisions to the fax machine before arriving at its modern form. Giovanni Caselli created the "pantelegraph", which became the first commercial fax link between Paris and Lyon in France around 1863. Based on Bain's ideas, Caselli's tall, cast-iron machine sent thousands of faxes annually.

181 Elevator (厢式电梯)

Evidence suggests that elevators existed in Ancient Rome in 336 BC, built by Archimedes. These were open cars and consisted of a platform with hoists. The first elevator to lift a passenger was designed for French King Louis in 1743.

In 1835, Frost and Stutt developed a counterbalance-type, traction method elevator called the "Teagle". In 1850, a hydraulic elevator working on steam was introduced. This was invented by Elisha G. Otis and called the "safety elevator". Otis solved many obstacles faced by earlier elevators.

The world's first passenger service elevator was installed in a five storey hotel on Broadway, New York, in 1857. Manufactured by the Otis Elevator Company, it was steam-powered, carried a maximum load of 450 kg and had a top speed of 12 m per minute. In 1867, Leon Edoux exhibited the hydraulic-power elevator at the Paris Exposition. With a top speed of 150 m per minute, it grew popular in Europe and the USA.

1. According to the passage, which elevator is safer and more advanced, the Teagle developed by Frost and Stutt or the hydraulic elevator invented by Elisha G. Otis? Give the reasons.
2. What's the difference between an elevator and an escalator?

182 General Anaesthesia（全身麻醉）

During the 1840s, three new methods of administering unconsciousness came about and were quickly used by doctors to improve their patients' condition. These were nitrous oxide[①], ether[②] and chloroform[③]. The ability of nitrous oxide to render people unconscious was recognised by English chemist Humphry Davy as early as 1789, who tried it on himself and realised that he didn't feel any pain under its influence. In the 1800s, American chemist Charles Jackson found a stronger anaesthetic — ether. Upon using ether, a person would lose consciousness as well as sensation. On 16 October, 1846, at Massachusetts General Hospital in Boston, the first public demonstration of ether anaesthesia was conducted by William Morton.

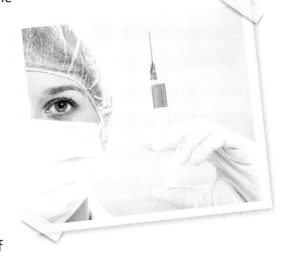

1. According to the passage, what were three new methods of administering unconsciousness?
2. Which was a stronger anaesthetic, nitrous oxide or ether? Give the reasons.

183 Antiseptics（抗菌剂）

In 1847, Ignaz Semmelweis, a Hungarian obstetrician at the Vienna General Hospital, was trying to find out why one among 30 young mothers was dying of puerperal fever in wards that were looked after by midwives. However, he realised that this number increased to almost one among five mothers in the wards looked after by medical students. Later, he realised that the students were reaching the wards after performing dissections on corpses and treating patients without cleaning and sanitising their hands. This caused infections among the patients. Semmelweis asked the students to wash their hands in diluted bleaching powder before operating. But this led to an increase in the death rate. In 1865, British surgeon Dr Joseph Lister began using carbolic acid[④] to clean his hands and tools. Later, this came to be used as an antiseptic because it did not lead to his patients' deaths.

1. According to the passage, what caused infections among the patients at the Vienna General Hospital in 1847?
2. What do you think of antiseptics? Give your reasons.

① nitrous oxide 氧化亚氮　② ether 乙醚　③ chloroform 氯仿　④ carbolic acid 苯酚

Possible Answers
（参考答案）

1
1. The oldest stone tools are known as the "Oldowan" toolkit. They include hammer stones, stone cores and sharp stone flakes.
2. Stone tools indicate the early culture and the abilities of human ancestors, so they play an important role in historical research.

2
1. Humans used it to generate light and heat, to clear forests for farming, to create ceramic objects out of clay and to aid the making of stone tools.
2. Fire may burn up our houses and even kill us.

3
1. They were made out of stone.
2. As technology improved, humans made more impressive blades as swords and slowly they were refined and made smaller. Eventually they were made the knives we use on the dining tables today.

4
1. When they travelled to Indonesia, crossing many leagues at sea.
2. I think they will disappear in the future because log rafts are not easy to control and they travel much more slowly than boats and ships./I don't think they will disappear in the future. Making log rafts can be a traditional skill and they are still widely used somewhere in China.

5
1. They used a wooden shaft as the handle of the spear; a hand-chiselled stone as the weapon and mixed adhesives to hold the spear together.
2. Yes, it is. I think it plays an important role in the history of humankind. Because before the spear was invented, humans used violence to acquire territories. When one group invented spears, the other group grew cautious and avoided fighting.

6
1. Because temperatures at that time changed drastically from very hot to very cold and humans had to find a way to protect themselves against the Arctic cold.
2. Fur has advantages and disadvantages. On the one hand, fur is a good choice to protect people against cold. On the other hand, many animals are killed for fur every year, which is quite cruel.

7
1. Around 8500 BC, during the Bronze Age.
2. Mud bricks refer to the bricks which are left in the sun to bake naturally, while kiln-fired bricks are the bricks which are fired with high temperature in the kiln.

8
1. It served as both a military weapon and a hunting tool.
2. I think it will disappear in the future, because it's more difficult and less efficient to use than other weapons like guns./I don't think it will disappear, because it represents a kind of tradition and nowadays shooting arrows with a bow has been a sport and even come into the Olympic Games.

9
1. Wild animals, abstract patterns and tracings of human hands.
2. Yes, I do. Since they recorded anything that occurred to humans at that time, we can get to know the life of ancient people from them. As a result, they gave us dear materials for further research into ancient people's living conditions.

10
1. It was made of the bone of a Griffon Vulture.
2. Yes, I think people in ancient China played the flute because the flute is around 43,000 years old and evidence of ancient flutes has

Possible Answers（参考答案）

been found in China.

11
1. The first version of the sewing needle was made of bone and it had a closed hook.
2. The earlier versions of the sewing needle were hook-shaped. The hooks might be damaged over time. The straight ones that we see today have an eye, so we can sew the thread through the eye directly and the sewing needle won't be damaged easily.

12
1. The oldest statue called "The Lion Man" and a statue called the "Venus of Hohle Fels".
2. The Chinese built the statue of the Leshan Giant Buddha because Buddhism was popular in China during Tang Dynasty. The statue expresses religious belief and possesses great cultural significance.

13
1. "The Spinner" was an ancient method used by rope makers. It involved tying a rock at the end of a stick and swinging it around to weave the rope.
2. Yes, they were strong because the ropes made of these materials were used by Egyptians for the construction of the Great Pyramids. And the Great Pyramids continue to stand strong even today.

14
1. To enhance their longevity and pigmentation.
2. The use of white lead in oil painting keeps the dried oil paint stable and it can be used with watercolour as well. That's why it is said to be one of the finest pigments ever made.

15
1. People held the meat above smoke that came from a fire. The smoke would help to dry out the meat and ward off bacteria.
2. No, it wouldn't. Because the sand would dehydrate the body and preserve the flesh.

16
1. Spruce root fibres, wild grass, stones and weights.
2. Some people use fishing rods to catch fish. Others use fish traps such as bottle traps and fish wheels.

17
1. Ancient humans lined their baskets with clay soil to fix the flaw that their vessels could not be used to hold water.
2. They drew the water with the basket lined with clay soil and left it aside. Furthermore, they dried the vessels in the hot sand or sun.

18
1. Two. They are oral language and written language.
2. No, they didn't. Most languages evolved naturally, but a few were invented. For example, the most successfully invented language is Esperanto, which originated in 1887.

19
1. Fermentation is a process which involves breaking down substances to create something new.
2. Yogurt, wine and cheese.

20
1. To keep insects from crawling into their mouths, ears or noses.
2. Cotton and plant seeds.

21
1. In the Mehrgarh region of Baluchistan, Pakistan.
2. It can stick construction materials together and make buildings stronger.

22
1. Since 6050 BC during the Neolithic era.
2. People can use salt to make glass, to melt ice on roads and to make soap.

23
1. Because it did not have a handle.
2. In ancient times, humans used axes to hunt animals and protect themselves. Nowadays, axes are used for carpentry, metallurgy and farming.

24
1. To measure the level of water and predict when the river would flood.
2. Humans used seismographs to predict earthquakes.

25
1. By using water and tree barks.
2. Leather is durable and flexible. It has excellent resistance to abrasion and wind.

26. 1. Tree sap.
    2. It is used for joining materials together, often for repairing broken handcrafts.
27. 1. As early as 2500 BC.
    2. The tortoise shell and the bamboo.
28. 1. Because they would help early humans walk and climb through rough terrain.
    2. Yes, I do. Because they can take the stress away from the toes and protect our feet.
29. 1. It is marked with lines indicating the hour of the day.
    2. Yes, I do. Because it told people the time by using shadows to record the position of the Sun in the sky in the past.
30. 1. Because they were formed of a solid, round discus.
    2. I think it is a very useful invention because it has made transportation easier and faster.
31. 1. In the 17th century.
    2. They can be used to hold tools and as a fashion statement.
32. 1. By tying planks of wood together and stuffing the gaps with dry grass, reeds and animal hide.
    2. They created a special ship which made the oarsmen to sit on two levels, one on top of the other and row at once.
33. 1. To help the saw cut better.
    2. A knife, a broken seashell and a shark's teeth.
34. 1. Cement, clay, gypsum and lime.
    2. Mixing volcanic ash or crushed brick into the cement mix.
35. 1. By melting copper and arsenic together.
    2. Because copper and tin could rarely be found underground together. In comparison, iron was easily available and more common.
36. 1. In the 13th century AD.
    2. I think they are useful because they can help people achieve the right look by fastening their clothing tightly.
37. 1. It is because of the triangular structure of the molecules in silk, which reflects light from several different angles.
    2. Yes, I have. It is an ancient trade route between China and the Mediterranean Sea, extending some 6,400 km and linking China with the Roman Empire.
38. 1. To make the bald heads look like they had hair and to protect the scalp from the heat.
    2. No, I don't think so. Compared to medical therapies for restoring hair, wigs are less expensive.
39. 1. They were made by sharpening the ends of bird feathers to make writing nibs.
    2. Eight kinds. They are thin twigs which men used to scratch on clay tablets, quills, quills with metal nibs, fountain pens, ballpoint pens, roller ball pens, felt pens and pens with ceramic nibs.
40. 1. Kohl is not just used as a cosmetic; it is a coolant and is also believed to protect children from evil eyes.
    2. Because women began to apply makeup more liberally than before.
41. 1. It looked like a long pole with a bucket or a bag attached to the end of it.
    2. They created the shadoof to refill the irrigation channels that they had built for the annual flooding.
42. 1. Because humans made great progress in building tools and implements in the Iron Age.
    2. Because the discovery of iron greatly promoted the development of tools and implements. Besides, iron plays an important role in the modern technology. It is used as the basis for creating materials able to withstand extreme temperatures, high pressures and so on.

Possible Answers（参考答案）

43
1. For keeping teeth and gums clean, whitening teeth and freshening breath.
2. It tells us the development of toothpastes.

44
1. Leonardo Da Vinci.
2. The electronic weighing scale is the most accurate one, since modern technology has developed fully enough to ensure its accuracy.

45
1. At the sites of Majiabang culture around Lake Taihu.
2. It tells us what a ploughshare is. It also tells us when and where ancient ploughshares were discovered.

46
1. During the 18th century.
2. We should brush our teeth regularly, have a healthy diet and go to the dentist's at least once a year.

47
1. They were between 3,000 and 3,300 years old.
2. With the rise of "mobile pastoralism" in Central Asia, nomads began moving their herds across the land on horseback and tunics and robes were not comfortable or suitable for long, bumpy rides as well as for battles.

48
1. It is 1,000 yards long and 60 feet wide.
2. Yes, I do. I live in Shanghai and Yongfu Road is such a historic road in Xuhui District which was built almost 100 years ago.

49
1. They were not quite systems at all. They were toilet-like cavities that drained to channels just outside.
2. It was a public bath and had a hole at one end which was used to drain the water out.

50
1. Concrete is a mixture of water, cement and certain other materials.
2. John Smeaton invented the hydraulic cement and Joseph Aspdin invented Portland cement.

51
1. Because humans felt the need of tongs to help them work with hot fires.
2. Surgical pliers, dental pliers, laboratory pliers and jewellery pliers.

52
1. The Egyptians.
2. Yes, I do. Three Gorges Dam and Gezhouba Dam.

53
1. Since 2800 BC.
2. Herbal leaves, essential oil and milk.

54
1. They were nothing more than holes in the ground.
2. It is mainly about the development of toilets.

55
1. Colours in the red, brown and orange families (are easily found in nature).
2. I think it is interesting because people dyed the fabric just by boiling it along with the source of the dye.

56
1. It is an architectural structure that distributes the weight of a heavy structure onto the ground below it.
2. In the Imperial Yuanmingyuan Garden, the Summer Palace and the Palace Museum.

57
1. During the Industrial Revolution of the 18th century.
2. Frying, boiling and steaming.

58
1. Because other civilisations had not discovered methods by which they could crush the seeds to extract oils.
2. No, they didn't. Because they thought it was a strange idea that cooking something in oil and fat would make the food tasty.

59
1. In 2000 BC.
2. I think they are inefficient and involve cruelty towards animals.

60
1. It began in the Mesopotamian era around 2000 BC.
2. She distilled flowers, calamus and oil. Then she filtered and mixed these ingredients with other aromatics and set them aside for a long time.

61
1. In the Bronze Age.
2. They used the arches for above-the-ground structures like bridges, aqueducts and gates.

62 1. They used their body parts for measuring things.
2. I think they are very useful because they can be used in carpentry, engineering and tailoring to measure distances or to rule straight lines.

63 1. By providing people additional protection from the elements of nature.
2. It's mainly about the history and the use of the umbrella.

64 1. The Zhou Dynasty.
2. Because they improved combat effectiveness.

65 1. Because the handles of its bronze blades were connected with flexible, curved, bronze spring.
2. I think scissors are very helpful because they can be used for cutting thin materials such as paper, cloth and ropes.

66 1. The people of ancient India discovered that by boiling the juice, it would reduce to form small, rocky particles.
2. Because it povides people with the heat, the nutrients and happiness.

67 1. It was used for making weapons.
2. Steel is used to make cars, planes and ships. Besides, it is widely used in the construction of roads and railways.

68 1. The Archimedes' screw was used to transfer water from low-lying water bodies to farms, to drain out land covered by sea water and to water the Hanging Gardens.
2. The passage is mainly about the uses of the Archimedes' screw and the evolution of it.

69 1. To wipe their necks and faces during summer.
2. Wool, pashmina and silk.

70 1. The Greeks.
2. I like taking a shower. Because it is more convenient and takes less time.

71 1. He, a shorthand writer, wore gloves during winter so that his work wouldn't get affected.

2. Gloves can be used as a fashion ornament, and medical gloves can block bacteria.

72 1. On the Isle of Crete.
2. The Romans used marble for tubs, and lead and bronze for pipes.

73 1. The first metal locks appeared towards the end of the first century.
2. I think it is helpful because it can prevent thieves from stealing valuables.

74 1. Calliper is a measuring instrument that resembles a compass for minute measurement.
2. It is mainly about the development of the calliper.

75 1. Because of a lack of maintenance and enemy attacks.
2. Yes, I do. Because aqueducts carried water from lakes and rivers to homes, public baths and even to farms, which brought convenience for the Romans. After the aqueducts fell out of use, the Roman population fell from a million to around 30,000.

76 1. Because he wanted to build a new weapon to strengthen his army.
2. The "ballista" was built to shoot arrows while the original catapult was built to shoot heavy loads.

77 1. It had a rounded shape, with straight top and bottom. And it had handles and in some cases, even wheels.
2. It is mainly about the appearance, advantages and origins of the barrel.

78 1. It could be used for starting fires and reading tiny letters.
2. Yes, I do. Because it can focus the sun's radiation to create a hot spot at the focus for fire starting.

79 1. It comes from an ancient Greek list of building supplies.
2. I think it is a wonderful invention because the wheelbarrow makes it more convenient

to carry loads.

80. 1. It uses a wheel, axle and a belt or rope to pull heavy loads with little force. The wheel is placed on the axle. It controls the movement and direction of the rope.
    2. The passage is mainly about how the pulley works and the history of it.

81. 1. It was used for irrigation and power generation.
    2. It's about the working principle of the waterwheel and the development of it.

82. 1. Plumbing is a drainage system that helps people empty waste water into sewers.
    2. I think it is of great importance. Plumbing has brought great convenience and cleanness to us, so that we could live in a cleaner and more civilized world.

83. 1. The pointer of a compass was made of lodestone, so it could orient itself in the north-south direction no matter how it was kept./ The needle of a compass is magnetised, so it could orient itself in the north-south direction no matter how it is kept.
    2. Adventurers. Because they may get lost in desserts or forests without compasses.

84. 1. It was decked in an auspicious shade of red, richly ornamented and gilded, and was equipped with red silk curtains to screen the bride.
    2. It is comfortable for passengers, but as it is powered by man, it's relatively slow compared with today's transports.

85. 1. Five. They are cinnamon, insects, nuts, seeds, eulachon or candlefish.
    2. Candles are used at birthday parties and romantic settings.

86. 1. Because when you push down on one end, the force of gravity also works with you, which saves you a lot of efforts.
    2. Egyptians used levers to lift up heavy stuff such as stones for there were no lifting machines at that time.

87. 1. They usually had large-scale domes, and strong base walls to hold up these domes.
    2. It suggests that domes can even be seen in remote areas.

88. 1. For easy consumption, construction, landfill or land reclamation activities.
    2. Extremely small part of an object.

89. 1. The calendar is based on exact astronomical observations of the longitude of the Sun and the phases of the Moon./The calendar is based on the cycle of the Moon as well as on Earth's course around the Sun.
    2. It embodies the great wisdom of ancient Chinese people.

90. 1. Ovens are commonly used for cooking, and they can also be used for manufacturing bricks and pottery.
    2. It is mainly about the development of ovens and how ovens work.

91. 1. When the Chinese encounterd the Romans and their postal system, the postal systems flourished.
    2. Though it is convenient for us to communicate with each other by phone, writing by hand is a better way to organize our ideas, improve our handwriting, and practise the skills of writing. Also, using hand-written pieces to reply to others is a way of showing respect.

92. 1. According to the passage, Americans' interest in travelling in groups to the beach for recreation caused the revolution.
    2. The swimsuit will continue to develop because of the increasing demands from people and the ever-changing fashion trend.

93. 1. Because people wrapped socks around the legs and feet.
    2. I think it's necessary to have different kinds of socks. First, they can satisfy people's demands for different uses. Second, they

also become a part of fashion.

94 1. Papyrus was used for writing in ancient Egypt and other Mediterranean cultures before paper was invented.
    2. No. Although the use of paper has been largely decreased due to the development of online office systems, some important files still need to be fixed in physical form.

95 1. Yes, we can.
    2. No. I'll be interested in it because it is part of our Chinese culture.

96 1. The toothbrush.
    2. Maybe it was invented by Doctor West, and it made teeth-brushing very comfortable.

97 1. It must have thick, supportive walls with limited gaps.
    2. Yes. Because they existed in ancient Egypt.

98 1. The cylindrical stem and a bulb weighted with mercury or lead inside the hydrometer.
    2. We can read the number on the scale inside the stem when the surface of the liquid touches the stem of the hydrometer.

99 1. He gathered documents detailing the locations of towns and augmented that information with the tales of travellers.
    2. I think this system was really useful because nowadays people are still using it.

100 1. To avoid carrying heavy and cumbersome metallic coins for transactions.
    2. I like using mobile payment systems because it's much easier and more convenient for people to take a mobile phone to travel to other places.

101 1. Because it has a truncated cone with an internal, cylindrical bore for holding an explosive charge and a projectile.
    2. I think it's a bad technology because it leads to more deaths./I think it's a good idea because it helps the development of civilisations.

102 1. It means 调料.
    2. Tangyuan/Glutinous rice dumplings. It is said to symbol a family reunion in China.

103 1. In China during the 13th century.
    2. Pistol, rifle and submachine gun.

104 1. It was a mixture of sulphur, saltpetre and charcoal.
    2. Gunpowder can be used to scare the enemies in a war. And it can also be useful in the field of building roads, mining and making fireworks.

105 1. Because the process of making velvet was fairly intricate.
    2. In the past, only those who had higher status in a society could afford this kind of clothing and it represented the position and power of a person in a community.

106 1. They were mainly used as weapons and fireworks.
    2. The story about a beauty named Chang'e flying to the Moon. It tells us that we Chinese people have been dreaming of travelling to space for thousands of years.

107 1. By finding out the amount of fabric she added to tunics.
    2. Yes, I do. I think the change and development of clothing goes hand in hand with the changes of people's life, concept and changes of times, technology. And of course it reflects the spread and communication of culture between different countries, thus making a society develop through time.

108 1. In *Fangyan*.
    2. Cleaned wool or cotton is first carded.

109 1. A spring-loaded mechanism released by the trigger.
    2. I believe those that are limited to fixed bursts of two, three or more rounds per squeeze are better because automatic ones may pose a threat to careless hunters or users and some

innocent people might be killed for no reason.

110 1. For more than 500 years.
     2. Yes. I do, because lacemaking made its appearance in both fashion and home decor. Without lacemaking, clothes or decorations won't be so pretty.

111 1. To correct people's vision if people cannot clearly see things that are at distance or even nearby.
     2. The history of the invention and the development of it.

112 1. To unscrew corks on wine and olive oil bottles.
     2. We wouldn't fasten things tightly and most of the machines might be loose and easily fall apart.

113 1. "Static electricity" can be created by rubbing fur with other objects, which causes them to attract each other.
     2. We can avoid it by not wearing clothes made of synthetic fiber.

114 1. Lead.
     2. To make it easy to be held./Because it was soft and brittle.

115 1. Because it was so heavy that it had to be held by a belt, which was worn around the waist.
     2. I don't think so. The watch is an art of sophistication and people wear it to show social status.

116 1. To make the block of type even.
     2. Because it has ensured that books, newspapers, magazines and other reading materials are produced in great numbers and it also plays an important role in promoting literacy among people.

117 1. It was called "bodies", "a stiff bodice" or "a pair of stays".
     2. It can do harm to people's body circulation if it is worn too tightly.

118 1. It was used exclusively for brewing tea.
     2. I like the authentic Yixing teapots better because they perfectly combine Chinese view of beauty and techniques of tea making.

119 1. It refers to the objective lens.
     2. In medical field. Because the bacteria and viruses are too tiny that we need microscopes to observe them.

120 1. A sign of status.
     2. Yes, it is. Because the shoes could offer stability./Because soldiers could use their bow and arrows more efficiently.

121 1. In 12th century Europe.
     2. I prefer knitted stockings because they offer more warmth.

122 1. Early projectiles were stone or metal objects that could fit down the barrel of the firearm.
     2. It's dangerous if it's used by wicked people.

123 1. They were used for protection and as a fashionable accessory. Additionally, they are markers of status, occupation and even political affiliation.
     2. Yes, I do. Because it can keep me cool in summer and warm in winter.

124 1. It was warmed by being held near steam and then rubbed across the fabric.
     2. Yes, I do. Because it can help remove wrinkles and create pleats.

125 1. The calendar was proposed by Aloysius Lilius, adopted by Pope Gregory XIII to correct the errors in the older Julian calendar and officially declared by him in 1582.
     2. In China, we both use Gregorian calendar and the traditional Chinese calendar—the lunisolar calendar, but I prefer the traditional Chinese calendar. It's because many important Chinese festivals are determined according to the lunisolar calendar and it even gives suggestions on agricultural productions.

126 1. The thermoscope was a thermometer without a scale and the thermoscope only showed the differences in temperatures.

2. They can show the differences in temperatures.

127 1. It's because it aided the later versions of locomotives.

2. The roads evolved from wooden wagonways to iron tramways, and the wagons evolved from the horse drawn wagons or carts to wagons with flanged wheels, then to the later versions of locomotives.

128 1. They used them to make fishing net floats, sandals and bottle stoppers.

2. Because the use of cork as a stopper grew wildly popular and people needed a lot of cork trees to produce corks, thus more people planted cork trees.

129 1. To hold the tops of their shirts in place at the collar.

2. He may wear a bow tie for the only accurate complement to such a suit is the bow tie.

130 1. Cornelius van Drebbel, a Dutch inventor, invented the first submarine.

2. Leonardo da Vinci drew sketches of a submarine, and later, William Bourne drew plans for it, and at last Cornelius van Drebbel created the first submarine.

131 1. It was called "water by fire".

2. In 1712, Thomas Newcomen invented a steam engine consisting of a piston or cylinder. From 1763 to 1775, James Watt improved Newcomen's engine, enabling factories to be sited away from rivers. In 1825, the Stockton & Darlington Railroad Company was the first railroad to carry both goods and passengers using locomotives designed by George Stephenson.

132 1. He used the telescope to see mountains and craters on the Moon. He also discovered sunspots on the Sun and Jupiter's moons.

2. Yes./No. I can use it to enjoy the Milky Way on clear nights.

133 1. They wear a tie for style.

2. Croatian soldiers wore ties to make it easy to identify themselves.

134 1. "Baros" means weight and "metron" means measure.

2. He used water to measure the air pressure to construct the first barometer.

135 1. It used whole blood.

2. Although people failed many times, their spirit of persistence helped them find the way to success./Blood tranfusion was a breakthrough that asked for great courage and efforts of exploration.

136 1. Leonardo da Vinci.

2. Maybe because people didn't have anything like a plane at that time but they still wanted to fly.

137 1. They used iceboxes to keep their food cool.

2. Yes, I do. Because it can prevent food from going bad in hot weather./No, I don't. Because old people may regard the refrigerator as a "safe". They may put food into the fridge for a long time and then eat it. This may cause health problems.

138 1. 96.

2. I think it's useful/helpful. It could spin 96 strands of yarn together and didn't require skilled operators, so it helped make yarn more efficiently.

139 1. They found circular razors made of bronze.

2. I want it to be more intelligent. For example, it can measure body temperature and heart rate while shaving.

140 1. It is an emulsion of oil, egg, lemon juice and/or vinegar along with different types of seasonings.

2. Because it is made of large amount of oil and eggs which have high calories.

Possible Answers（参考答案）

141 1. Yes, it is.
2. It's 3N.

142 1. Because he developed a process to manufacture carbonated mineral water using the same process discovered by Joseph Priestley.
2. It is refreshing and has diverse flavours.

143 1. He patented a 16-spindle spinning jenny.
2. Because it was the first improvement on the spinning wheel and the invention of it paved the way for the Industrial Revolution.

144 1. During the first century BC.
2. Because people are very busy and do not want to waste too much time on meals and sandwiches are quick and easy to perpare.

145 1. Because his patient didn't contract smallpox, even when he was deliberately exposed to variola.
2. Yes, I do. Although being a scientist like Edward is dangerous, I want to make some contributions to the human being.

146 1. Yes, it was.
2. "Palmipède" had a small, wood-fired steam engine, while "Pyroscaphe" had a Newcomen steam engine.

147 1. No, it didn't.
2. No, they couldn't. Because the writer mentioned that the condition of the roads around the bridge improved in the years after its construction.

148 1. It stayed aloft for almost four minutes.
2. Because people at that time were full of spirits of challenge, and the teacher loved new inventions.

149 1. Threshers can separate grains from stalks and husks more easily and less time-consuming.
2. Wheat, rice and oats.

150 1. Because Whitney's cotton gin successfully pulled out seeds from cotton bolls.
2. Because the invention made cotton producing boom, and it changed the relationship between the South and the North, which influenced the Civil War.

151 1. A circle gives the least amount of continuous contact with its surroundings.
2. Because friction can generate enormous amounts of heat at the speed at which many bearings operate.

152 1. Alessandro Volta.
2. Yes. It is very dangerous because of the experiment where an electrical wire was placed in a jar filled with methane gas, when an electrical spark was sent through the wire, the jar exploded.

153 1. To plot the position of a ship on navigational charts.
2. In my opinion, it is still necessary for students to learn how to use protractors because we can learn the basic method and principles to measure angles while using them.

154 1. Both Sir George Cayley and Otto Lilienthal.
2. A hang glider is devised to carry a human passenger who is suspended beneath the sail in the air. As they are usually launched from a high point and drift slowly, the passengers can see marvelous views and have amazing experience, so hang gliders can be used as a kind of entertainment.

155 1. They discovered that quinine cured malaria and they extracted quinine from the bark of the South American cinchona tree.
2. Yes, I do. Artemisinin is also a substance extracted from plants to cure malaria.

156 1. Four stages.
(1) Frenchman Philippe de Girard passed the idea to a British merchant called Peter Durand, and patented the tin can in 1810.
(2) In 1813, John Hall and Bryan Dorkin opened the first commercial canning factory in England.
(3) In 1846, Henry Evans invented the

machine that manufactured 60 tin cans per hour.

(4) In 1847, Allen Taylor patented his machine-stamped method of producing tin cans.

2. It took 37 years.

157 1. For 117 years.

2. While experimenting with a solid electrode in an electrolyte solution, the phenomenon that the voltage developed when light fell on the electrode gave him the idea .

158 1. Steam-powered engines.

2. His experience and knowledge about how to operate steam-powered engines.

159 1. It was quick and efficient. It harvested more wheat and other grains than manual cutting. It reduced labour costs and the danger of weather destroying crops.

2. Because it manufactured many machines with advanced performance and exported them to other countries.

160 1. Because he couldn't hear anything when he tapped his hand on the patient's back, for the patient was a plump woman, and he didn't want to put his ear on her chest.

2. Yes, it helps doctors listen to the patients' heartbeat more easily and clearly.

161 1. He observed that a pinhole can form an inverted and focused image, which was the principle of early cameras.

2. Many people contributed a lot to the invention of cameras.

162 1. Since it didn't have any petals, one had to propel the bicycle forward using their feet.

2. Around 1790—Celerifere.

In 1816—Draisienne.

Between 1830s to 1840s—The first bicycle with foot pedals.

In the 1860s—An improved bicycle.

During the 1880s—A chain with sprockets and air-filled tyres were added to the bicycles.

In the 1970s—The gear system was added.

163 1. Braces used leather loops that were attached to buttons on the pants, while suspenders use mental clasps that clasped to the trousers's waistband.

2. Yes, I do. Because when we wear suspenders, we don't need to use belts to hold up the trousers.

164 1. By striking the matchstick against the rough surface of the matchbox.

2. Because they have a strong smell and don't work if stored in a humid environment.

165 1. By using rubber dissolved in coal-tar naphtha for cementing two pieces of cloth together.

2. No, it wouldn't. Because vulcanised rubber was invented and Macintosh's fabric improved as it could withstand temperature changes.

166 1. No, it didn't. Because the driver would pick up or set down passengers anywhere upon their request.

2. I would say, "Thank you so much. It was really convenient to take the bus".

167 1. He injured his eye while he was playing. Though he was offered the best medical attention, an infection developed and spread to his other eye, leading to the blindness in both eyes.

2. I can see it on the buttons in the lift. It is essential for those who are visually deficient.

168 1. No, he didn't receive the recognition because of the botched filling of his patent at the patent office.

2. Yes, I would. Because his idea made a contribution towards the modern sewing machine.

169 1. They turned the skills of horse travel into those for train travel.

2. The development of the public transporta-

tion.

170 1. Because it "combines" the job of the header and the thresher.

2. Yes, I would. Because it made farming safer, more profitable and helped food reach many people.

171 1. Because the telephone, fax machine and Internet replaced it.

2. Yes, I do. Because it sounds interesting and is a bit like a vintage art of techinque.

172 1. The optical telegraph used semaphore and depended on a line of sight for communication.

2. The non-electric telegraph invented by Claude Chappe, the crude telegraph invented by Samuel Soemmering, the electromechanical telegraph invented by Samuel Morse.

173 1. It's a mail system that delivered letters and small parcels within London for one penny.

2. They are used to confirm that the postage for the parcel or letter is prepaid.

174 1. William Hamilton Merritt visualised it, but he didn't build it at last.

2. It's an inexpensive solution to the problem of long spans over navigable streams or at other sites where it is difficult to build piers. But the bridge may not be stable or strong enough against wind forces and heavy loads, etc.

175 1. The increase in the use of paper during the 19th century created the need for an efficient paper fastener and caused people to design it.

2. The stapler created in 1879 and called "McGill's Patent Single Stroke Staple Press".

176 1. In 1825.

2. Nowadays people use aluminium to make containers which are difficult to get rusty, such as aluminium cans.

177 1. It's used to measure the difference in electrical potential between two points of an electric circuit.

2. Yes, I have. I once used it in a physics class.

178 1. A lighter-than-air gas is filled in the dirigible, because it can provide lift.

2. Because the plane replaced it for a faster and safer flight.

179 1. It was the first pin to have a clasp that prevented the sharp edge from poking someone.

2. He didn't invent the safety pin on purpose. Because when he was thinking about a way to pay a 15-dollar debt, he twisted the wire for three hours and then realised that he had created something different and useful.

180 1. Because his invention allowed a clearer image through better synchronisation./His image telegraph was successful.

2. When we are choosing a fax machine, we should consider the synchronisation, clearness of the images and the efficiency of the machine.

181 1. The hydraulic elevator invented by Elisha G. Otis. Because Otis solved many obstacles faced by earlier elevators.

2. An elevator goes up and down vertically and carries people from one floor to another floor while an escalator is a set of moving stairs that moves between only two floors.

182 1. Nitrous oxide, ether and chloroform.

2. Ether. Because upon using ether, a person would lose consciousness as well as sensation while the ability of nitrous oxide was to render people unconscious.

183 1. The medical students reached the wards after performing dissections on corpses and treated patients without cleaning and sanitising their hands.

2. I think they are helpful because they can help protect patients from infections.

Words and Expressions
（生词与短语）

1

theory	/ˈθɪəri/	n.	理论；原理
natural selection	/ˈnætʃrəl sɪˈlekʃn/	n.	自然选择；物竞天择
evolve	/ɪˈvɒlv/	v.	进化；演化
ape-like	/eɪp laɪk/	adj.	类猿的
ancestor	/ˈænsestə(r)/	n.	祖先；祖宗
intellectual	/ˌɪntəˈlektʃuəl/	adj.	智力的；聪明的
differentiate	/ˌdɪfəˈrenʃieɪt/	v.	区分，区别
era	/ˈɪərə/	n.	时代；年代；纪元
toolkit	/ˈtuːlkɪt/	n.	工具包，工具箱
Oldowan	/ˈɒldəwən/	adj.	奥尔杜韦文化的
core	/kɔː(r)/	n.	（物体的）中心部分
flake	/fleɪk/	n.	薄片

2

crucial	/ˈkruːʃl/	adj.	至关重要的；关键性的
evolution	/ˌevəˈluːʃn/	n.	演变；进化论；进展
generate	/ˈdʒenəreɪt/	v.	产生；引起
ceramic	/səˈræmɪk/	adj.	陶器的；制陶的
evidence	/ˈevɪdəns/	n.	证据；证明
archaeologist	/ˌɑːkiˈɒlədʒɪst/	n.	考古学家
unearth	/ʌnˈɜːθ/	v.	挖掘；找到；使出土
charred	/tʃɑːd/	adj.	烧焦的；烧黑的
estimate	/ˈestɪmeɪt/	v.	估计；估算
dispute	/ˈdɪspjuːt/	v.	争论；争执；辩论
state	/steɪt/	v.	说明；陈述；声明
date back to			追溯到；始于

3

eventually	/ɪˈventʃuəli/	adv.	最后；终于
obtain	/əbˈteɪn/	v.	（尤指经努力）获得
crude	/kruːd/	adj.	粗糙的
rudimentary	/ˌruːdɪˈmentri/	adj.	基础的；基本的；原始的
blade	/bleɪd/	n.	刀片
refine	/rɪˈfaɪn/	v.	改进；改善

4

log	/lɒg/	n.	原木

migrate	/maɪˈɡreɪt/	v.	迁移；移居
league	/liːɡ/	n.	里路（长度单位，约等于3英里或4,000米）
region	/ˈriːdʒən/	n.	区域
back then			那时；过去

5

spear	/spɪə(r)/	n.	矛
feat	/fiːt/	n.	功绩；技艺
engineer	/ˌendʒɪˈnɪə(r)/	v.	设计制造
species	/ˈspiːʃiːz/	n.	物种
shaft	/ʃɑːft/	n.	杆，柄
hand-chiselled	/hænd ˈtʃɪzld/	adj.	手凿的
adhesive	/ədˈhiːsɪv/	n.	黏合剂；黏着剂
testimony	/ˈtestɪməni/	n.	证据；证明
complex	/kəmˈpleks/	adj.	复杂的
reasoning	/ˈriːzənɪŋ/	n.	推理；论证
acquire	/əˈkwaɪə(r)/	v.	获得
territory	/ˈterətəri/	n.	领土；领地；地盘
violence	/ˈvaɪələns/	n.	暴力；暴行
cautious	/ˈkɔːʃəs/	adj.	小心的；谨慎的
tension	/ˈtenʃn/	n.	紧张局势（或关系、状况）
result in			导致；结果是；造成

6

primitive	/ˈprɪmətɪv/	adj.	原始的；远古的
trendy	/ˈtrendi/	adj.	时髦的；赶时髦的
controversial	/ˌkɒntrəˈvɜːʃl/	adj.	引起争论的；有争议的
volatile	/ˈvɒlətaɪl/	adj.	易变的；无定性的
drastically	/ˈdrɑːstɪkli/	adv.	大幅度地；大大地
Arctic	/ˈɑːktɪk/	adj.	北极的；北极地区的；极冷的
mammoth	/ˈmæməθ/	n.	猛犸（象）
musk ox	/mʌsk ˈɒks/	n.	（北极）麝牛［ox的复数是oxen］

7

brick	/brɪk/	n.	砖；砖块
civilisation	/ˌsɪvəlaɪˈzeɪʃn/	n.	文明；文化
straw	/strɔː/	n.	稻草
mould	/məʊld/	n.	模型；模具；模子
Mesopotamian	/ˌmesəpəˈteɪmiən/	adj.	美索不达米亚的
kiln	/kɪln/	n.	窑
masonry	/ˈmeɪsənri/	n.	砌筑；砌砖；砌石
shape up			发展；成形

Words and Expressions
（生词与短语）

kiln-fired brick			窑烧砖

8

bow	/bəʊ/	n.	弓
arrow	/'ærəʊ/	n.	箭头；箭号
formidable	/fɔː'mɪdəbl/	adj.	可怕的；令人敬畏的
cave	/keɪv/	n.	洞穴
military	/'mɪlətri/	adj.	军事的；军队的
Mediterranean	/ˌmedɪtə'reɪniən/	adj.	地中海的
Eurasian	/juə'reɪʒn/	adj.	欧亚的
crossbow	/'krɒsbəʊ/	n.	十字弓；弩；弩弓
longbow	/'lɒŋbəʊ/	n.	长弓；大弓
mounted	/'maʊntɪd/	adj.	骑马的
archer	/'ɑːtʃə(r)/	n.	弓箭手
recurved	/rɪ'kɜːvd/	adj.	后弯的；内弯的
horn	/hɔːn/	n.	（羊、牛等动物的）角
primarily	/praɪ'merəli/	adv.	主要地；根本地
for instance			例如
come up with			想出（一个主意或计划）；提出；提供

9

phase	/feɪz/	n.	阶段；时期
originate	/ə'rɪdʒəneɪt/	v.	起源；发源；发端于
apex	/'eɪpeks/	n.	顶点；最高点
tribe	/traɪb/	n.	部落；族
bison	/'baɪsn/	n.	野牛
aurochs	/'ɔːrɒks/	n.	原牛（已灭绝的古代野牛）
abstract	/æb'strækt/	adj.	抽象（派）的
pattern	/'pætən/	n.	图案；花样；式样
tracing	/'treɪsɪŋ/	n.	描摹；描图；摹图
retire	/rɪ'taɪə(r)/	v.	离开（尤指去僻静处）
occur	/ə'kɜː(r)/	v.	发生；出现；存在于；出现在
vast	/vɑːst/	adj.	大量的
palette	/'pælət/	n.	（画家使用的）主要色彩
option	/'ɒpʃn/	n.	可选择的事物；选择
sketch	/sketʃ/	v.	画素描；画速写
prehistoric	/ˌpriːhɪ'stɒrɪk/	adj.	史前的；有文字记载以前的；远古的
scrape	/skreɪp/	v.	刮掉；削去

10

instrument	/'ɪnstrʊmənt/	n.	仪器；工具；器械
boxwood	/'bɒkswʊd/	n.	（植）黄杨树；黄杨木材

| brass | /brɑːs/ | n. | 黄铜 |

11

version	/'vɜːʃn/	n.	型式；版本
thread	/θred/	n.	（棉、毛、丝等的）线
hook	/hʊk/	n.	钩；挂钩
ivory	/'aɪvəri/	n.	象牙；（某些其他动物的）长牙
sewing needle			（缝纫用）针

12

statue	/'stætjuː/	n.	雕像；塑像；铸像
possess	/pə'zes/	v.	具有（特质）
significance	/sɪɡ'nɪfɪkəns/	n.	（尤指对将来有影响的）重要性，意义

13

twist	/twɪst/	v.	捻；扭转
braid	/breɪd/	v.	编织
swing	/swɪŋ/	v.	（使）弧线运动，转动
weave	/wiːv/	v.	（用手或机器）编，织
lay the foundation			奠定基础
water reed			水芦
grass leather			草革
hand tools			手工工具

14

pigment	/'pɪɡmənt/	n.	色素；颜料
minimal	/'mɪnɪməl/	adj.	极小的；极少的
purify	/'pjʊərɪfaɪ/	v.	净化；提纯
crush	/krʌʃ/	v.	压碎；挤压变形
enhance	/ɪn'hɑːns/	v.	提高；增强
longevity	/lɒn'dʒevəti/	n.	持久
pigmentation	/ˌpɪɡmen'teɪʃn/	n.	色素沉着
smalt	/smɔːlt/	n.	大青色
cobalt	/'kəʊbɔːlt/	n.	深蓝色；钴蓝
vermillion	/və'mɪljən/	n.	银朱；朱红色
contribution	/ˌkɒntrɪ'bjuːʃn/	n.	贡献

15

preservative	/prɪ'zɜːvətɪv/	n.	防腐剂；保护剂
perish	/'perɪʃ/	v.	腐烂
preservation	/ˌprezə'veɪʃn/	n.	保存；保护；防腐
bacteria	/bæk'tɪəriə/	n.	细菌［单数是bacterium］
spice	/spaɪs/	n.	（调味）香料
vinegar	/'vɪnɪɡə(r)/	n.	醋

Words and Expressions
（生词与短语）

bury	/ˈberi/	v.	埋葬；埋藏；安葬
dehydrate	/diːˈhaɪdreɪt/	v.	使（身体）脱水
preserve	/prɪˈzɜːv/	v.	保留；保存；保鲜
flesh	/fleʃ/	n.	肉体
microorganism	/ˌmaɪkrəʊˈɔːɡənɪzəm/	n.	微生物
dry out			变干
ward off			挡开；架开
keep ... at bay			围住；不使……接近

16

grab	/ɡræb/	v.	抓住；攫取
stab	/stæb/	v.	刺；戳；捅
spruce	/spruːs/	n.	云杉
fibre	/ˈfaɪbə(r)/	n.	纤维
willow	/ˈwɪləʊ/	n.	柳树
carving	/ˈkɑː(r)vɪŋ/	n.	雕刻品；雕刻图案
literature	/ˈlɪtrətʃə(r)/	n.	文学；文学作品
feel like doing			想要做某事
rock carving			石雕
hint at			暗示；示意

17

pottery	/ˈpɒtəri/	n.	陶器（尤指手工制的）；陶土
vessel	/ˈvesl/	n.	器皿
flaw	/flɔː/	n.	瑕疵
line	/laɪn/	v.	（用……）做衬里
leftover	/ˈleftəʊvə(r)/	adj.	剩余的
shrink	/ʃrɪŋk/	v.	收缩
furthermore	/ˌfɜːðəˈmɔː(r)/	adv.	此外；而且；再者
clay soil			粘土
leave aside			搁置一边
get soaked into			浸泡进入
dry up			干涸

18

stock	/stɒk/	n.	储备；牲畜
stylus	/ˈstaɪləs/	n.	描画针；记录针
millennium	/mɪˈleniəm/	n.	一千年
obsolete	/ˈɒbsəliːt/	adj.	淘汰的；废弃的；过时的
decipher	/dɪˈsaɪfə(r)/	v.	破译
attempt	/əˈtempt/	n.	尝试；企图；努力
Polish	/ˈpɒlɪʃ/	adj.	波兰的

trace back to			追溯到
tally mark			计数符号
keep a count of			计算
the passage of time			时间的推移
cuneiform script			楔形文字
clay tablet			黏土平板

19

fermentation	/ˌfɜːmenˈteɪʃn/	n.	发酵
involve	/ɪnˈvɒlv/	v.	包含；需要；使成为必然部分（或结果）
substance	/ˈsʌbstəns/	n.	物质；物品；东西
alcoholic	/ˌælkəˈhɒlɪk/	adj.	酒精的；含酒精的
beverage	/ˈbevərɪdʒ/	n.	（除水以外的）饮料
ferment	/fəˈment/	v.	（使）发酵
alcohol	/ˈælkəhɒl/	n.	酒精
associate	/əˈsəʊʃieɪt/	v.	联想；联系
yeast	/jiːst/	n.	酵母；酵母菌
zymologist	/zaɪˈmɒlədʒɪst/	n.	发酵学家；酶学家
define	/dɪˈfaɪn/	v.	解释（词语）的含义；给（词语）下定义
respiration	/ˌrespəˈreɪʃn/	n.	呼吸
break down			使分解（为）；使变化（成）

20

pillow	/ˈpɪləʊ/	n.	枕头
fluffy	/ˈflʌfi/	adj.	松软的
carve	/kɑːv/	v.	雕刻
crawl	/krɔːl/	v.	（昆虫）爬行
inspire	/ɪnˈspaɪə(r)/	v.	赋予灵感；引起联想；启发思考
elaborately	/ɪˈlæbərətli/	adv.	精巧地；苦心经营地
stuff	/stʌf/	v.	填满；装满；塞满
keep off			使……不接近（或远离）

21

mortar	/ˈmɔːtə(r)/	n.	灰泥；砂浆
paste	/peɪst/	n.	糨糊
abundance	/əˈbʌndəns/	n.	大量；丰盛；充裕

22

refined	/rɪˈfaɪnd/	adj.	精炼的；提纯的；精制的
currency	/ˈkʌrənsi/	n.	通货；货币
extract	/ɪkˈstrækt/	v.	提取；提炼
immense	/ɪˈmens/	adj.	极大的；巨大的
barter	/ˈbɑːtə(r)/	v.	（同某人）以物易物；以（财产或劳务等）作交换

Words and Expressions
（生词与短语）

| slab | /slæb/ | n. | 厚片；厚块 |
| fight over | | | 为……争夺；为……争吵 |

23

axe	/æks/	n.	斧
attach	/əˈtætʃ/	v.	把……固定，把……附（在…上）
hide	/haɪd/	n.	（尤指买卖或用作皮革的）皮，毛皮
wind	/waɪnd/	v.	缠绕［过去式和过去分词是wound］
dawn	/dɔːn/	n.	开端；曙光；萌芽
copper	/ˈkɒpə(r)/	n.	铜
replicate	/ˈreplɪkeɪt/	v.	复制；（精确地）仿制
carpentry	/ˈkɑːpəntri/	n.	木工；木工工艺；木匠活
metallurgy	/məˈtælədʒi/	n.	冶金
a length of			一段；一截
on a large scale			大规模地

24

irrigation	/ˌɪrɪˈɡeɪʃn/	n.	灌溉
concept	/ˈkɒnsept/	n.	概念；观念
channel	/ˈtʃænl/	v.	（经过通道）输送，传送
overflow	/ˌəʊvəˈfləʊ/	v.	漫出；溢出
devise	/dɪˈvaɪz/	v.	设计；制定；发明；创造
nilometer	/naɪˈlɒmɪtə/	n.	水位计
divert	/daɪˈvɜːt/	v.	使转向；使绕道；转移
via	/ˈvaɪə; ˈviːə/	prep.	通过，凭借（某人、系统等）
canal	/kəˈnæl/	n.	运河；灌溉渠
dam	/dæm/	n.	水坝；拦河坝

25

tomb	/tuːm/	n.	坟墓；冢
indicate	/ˈɪndɪkeɪt/	v.	象征；暗示
sandal	/ˈsændl/	n.	凉鞋
bucket	/ˈbʌkɪt/	n.	（有提梁的）桶；吊桶
shroud	/ʃraʊd/	n.	裹尸布；寿衣
rot	/rɒt/	v.	（使）腐烂
stiffen	/ˈstɪfn/	v.	（使）难以弯曲，发僵
initial	/ɪˈnɪʃl/	adj.	最初的；开始的；第一的
tanning	/ˈtænɪŋ/	n.	制革（法）
durable	/ˈdjʊərəbl/	adj.	耐用的；持久的
flexible	/ˈfleksəbl/	adj.	柔韧的；可弯曲的；有弹性的
rub	/rʌb/	v.	涂；抹；搽
expose	/ɪkˈspəʊz/	v.	暴露；显露；露出
contribute	/kənˈtrɪbjuːt/	v.	是……的原因之一

135

formula	/ˈfɔːmjələ/	n.	方案；方法
bark	/bɑː(r)k/	n.	树皮
full-fledged	/ˌfʊlˈfledʒd/	adj.	成熟的；完全合格的
tannery	/ˈtænəri/	n.	鞣皮厂；皮革厂
medieval	/ˌmediˈiːvl/	adj.	中世纪的（约公元1000到1450年）
be considered to be			被视为
start out as			起初是
set up			创建；建立；开办

26

sap	/sæp/	n.	（植物体内运送养分的）液，汁
grain	/ɡreɪn/	n.	谷物
tar	/tɑː(r)/	n.	焦油；沥青；柏油
beeswax	/ˈbiːzwæks/	n.	蜂蜡；黄蜡
path-breaking	/ˈpæθˌbreɪkɪŋ/	adj.	开路先锋的；开创性的
decade	/ˈdekeɪd/	n.	十年，十年期（尤指一个年代）
egg whites			蛋清

27

papyrus	/pəˈpaɪrəs/	n.	纸莎草
inscribe	/ɪnˈskraɪb/	v.	在……上写（词语、名字等）；题；刻
scroll	/skrəʊl/	n.	（供书写的）长卷纸，卷轴
document	/ˈdɒkjumənt/	n.	文件；文献；证件
loose	/luːs/	adj.	未固定牢的；可分开的
codex	/ˈkəʊdeks/	n.	古书手抄本
religious	/rɪˈlɪdʒəs/	adj.	宗教信仰的；宗教的
literary	/ˈlɪtərəri/	adj.	文学的；文学上的
scripture	/ˈskrɪptʃə(r)/	n.	（某宗教的）圣典，经文

28

perishable	/ˈperɪʃəbl/	adj.	易腐烂的；易变质的
terrain	/təˈreɪn/	n.	地形；地带；地势
observation	/ˌɒbzəˈveɪʃn/	n.	观察；观测；监视

29

sundial	/ˈsʌndaɪəl/	n.	日晷
horizontal	/ˌhɒrɪˈzɒntl/	adj.	水平的；地平线的
derive	/dɪˈraɪv/	v.	追寻起源；推究；由来
cast	/kɑːst/	v.	投射（光、影等）
derive from			来自；起源于

30

| contraption | /kənˈtræpʃn/ | n. | 奇异的机械；奇特的装置 |
| vehicle | /ˈviːəkl/ | n. | 车辆；工具；交通工具 |

Words and Expressions
（生词与短语）

align	/əˈlaɪn/	v.	排列整齐；使对齐
axle	/ˈæksl/	n.	车轴
snugly	/ˈsnʌgli/	adv.	紧贴地；舒适地
potter	/ˈpɒtə(r)/	n.	陶工
discus	/ˈdɪskəs/	n.	铁饼
revolutionise	/ˌrevəˈluːʃənaɪz/	v.	彻底改变；完全变革
modification	/ˌmɒdɪfɪˈkeɪʃn/	n.	改进；修改；改变
significant	/sɪgˈnɪfɪkənt/	adj.	有重大意义的；显著的
spoked wheel			辐条式车轮
wire-spoked wheel			线辐轮

31

snug	/snʌg/	adj.	贴身的；紧身的；严密的；严实的
statement	/ˈsteɪtmənt/	n.	说明；说法；表白；表态
practical	/ˈpræktɪkl/	adj.	实际的；真实的；客观存在的；切实可行的
accessory	/əkˈsesəri/	n.	饰品；附属品
braided	/ˈbreɪdɪd/	adj.	以饰带装饰的
studded	/ˈstʌdɪd/	adj.	用饰钉装饰的
jewelled	/ˈdʒuːəld/	adj.	镶有宝石的；带首饰的
utility	/juːˈtɪləti/	adj.	多用途的；多效用的；多功能的
iconic	/aɪˈkɒnɪk/	adj.	符号的；图标的；图符的；偶像的
component	/kəmˈpəʊnənt/	n.	成分；部件；组成部分
fashion statement			时尚宣言

32

waterproof	/ˈwɔːtəpruːf/	adj.	不透水的；防水的；耐水的
sail	/seɪl/	n.	帆
oar	/ɔː(r)/	n.	桨；船桨
bireme	/ˈbaɪriːm/	n.	（古代的）对排桨海船
innovate	/ˈɪnəveɪt/	v.	创新；引入（新事物、思想或方法）；改革
hull	/hʌl/	n.	船体；船身
element	/ˈelɪmənt/	n.	要素；原理
counterpart	/ˈkaʊntəpɑːt/	n.	对应的事物

33

saw	/sɔː/	n.	锯
mythology	/mɪˈθɒlədʒi/	n.	（统称）神话；某文化（或社会等）的神话
jaw	/dʒɔː/	n.	颌
archaeology	/ˌɑːkiˈɒlədʒi/	n.	考古学
jagged	/ˈdʒægɪd/	adj.	凹凸不平的；有尖突的；锯齿状的

34

| cement | /səˈment/ | n. | 水泥 |

137

hurdle	/ˈhɜːdl/	n.	障碍；难关
recipe	/ˈresəpi/	n.	方法；诀窍
hydraulic	/haɪˈdrɔːlɪk/	adj.	（通过水管等）液压的；液压驱动的
volcanic	/vɒlˈkænɪk/	adj.	火山的；火山引起的
ash	/æʃ/	n.	灰；灰烬
binding substance			粘合物质
cement mix			水泥配料

35

alloy	/ˈælɔɪ/	n.	合金
formula	/ˈfɔːmjələ/	n.	配方
obstacle	/ˈɒbstəkəl/	n.	障碍；阻碍
fade	/feɪd/	v.	变淡，变暗；逐渐消逝
pave	/peɪv/	v.	（用砖石）铺（地）
fade away			逐渐消失
pave the path for			为……铺平道路

36

curved	/kɜːvd/	adj.	呈弯曲状的；弧形的
loop	/luːp/	n.	环形；环状物；圆圈
fabric	/ˈfæbrɪk/	n.	织物；布料
debt	/det/	n.	债务；欠款
pluck	/plʌk/	v.	摘；拔；采摘；拔掉
fit into			（使）适合；与……融为一体
pluck off			扯去

37

fibre	/ˈfaɪbə(r)/	n.	（织物或绳等）纤维制品
dissolve	/dɪˈzɒlv/	v.	使（固体）溶解
shimmering	/ˈʃɪmərɪŋ/	adj.	闪烁的；微微发亮的
molecule	/ˈmɒlɪkjuːl/	n.	分子
legend	/ˈledʒənd/	n.	传说；传奇故事
mulberry	/ˈmʌlbəri/	n.	桑树；桑葚
sip	/sɪp/	v.	小口喝；抿
unravel	/ʌnˈrævl/	v.	松开
realise	/ˈriːəlaɪz/	v.	认识到，明白
penalty	/ˈpenəlti/	n.	处罚；刑罚；惩罚
reveal	/rɪˈviːl/	v.	透露；揭示
constant	/ˈkɒnstənt/	adj.	连续发生的；不断的；重复的
immigrant	/ˈɪmɪɡrənt/	n.	（外来）移民；外侨
headgear	/ˈhedɡɪə(r)/	n.	帽子；头戴之物
despite	/dɪˈspaɪt/	prep.	尽管；即使

silkworm cocoons			蚕茧
legend has it that			传说；据传说；传奇的是
fall into			落入
pull out			拔出；拉出

38

wig	/wɪg/	n.	假发
shave	/ʃeɪv/	v.	剃（须发）
dual	/'djuːəl/	adj.	两部分的；双重的
scalp	/skælp/	n.	头皮
distinct	/dɪ'stɪŋkt/	adj.	清晰的；明显的
elaborate	/ɪ'læbəreɪt/	adj.	精心制作的
adorn	/ə'dɔːn/	v.	装饰；装扮
palm	/pɑːm/	n.	棕榈树
upper class			上等阶层
lower class			下等阶层

39

twig	/twɪg/	n.	嫩枝；细枝；小枝
scratch	/skrætʃ/	v.	刮出痕迹
hollow	/'hɒləʊ/	adj.	中空的；空心的
quill	/kwɪl/	n.	翎笔；羽管笔
nib	/nɪb/	n.	钢笔尖
eliminate	/ɪ'lɪmɪneɪt/	v.	消除；消灭
ceramic	/sə'ræmɪk/	n.	陶瓷制品
fountain pen			自来水笔
ballpoint pen			圆珠笔
roller ball pen			水性笔；中性笔；钢珠笔
felt pen			毡头笔

40

kohl	/kəʊl/	n.	眼线
cosmetic	/kɒz'metɪk/	n.	化妆品；美容品
coolant	/'kuːlənt/	n.	冷却剂；散热剂
evil	/'iːvl/	adj.	恶毒的；邪恶的；有害的
dip	/dɪp/	v.	蘸；浸
muslin	/'mʌzlɪn/	n.	细平布（旧时尤用于做衣物和窗帘）
sandalwood	/'sændlwʊd/	n.	檀香油（提取自檀香木，用于制作香水）
wick	/wɪk/	n.	灯芯；烛芯
soot	/sʊt/	n.	油烟；煤烟子
liberally	/'lɪbərəli/	adv.	大方地；自由地
dip into			浸在……里

41

shadoof	/ʃɑːˈduːf/	n.	（中东地区农民用的）汲水吊杆；桔槔
crane-like	/kreɪn laɪk/	adj.	类似起重机的
lever	/ˈliːvə(r)/	n.	杠杆
mechanism	/ˈmekəˌnɪz(ə)m/	n.	机制；机械装置
refill	/riːˈfɪl/	v.	再装满，重新装满
channel	/ˈtʃænl/	n.	水渠；沟渠；河槽
annual	/ˈænjuəl/	adj.	每年的；一年一次的；年度的；一年的

42

versatile	/ˈvɜːsətaɪl/	adj.	多用途的；多功能的
implement	/ˈɪmplɪmənt/	n.	工具，器具（常指简单的户外用具）
undergo	/ˌʌndəˈɡəʊ/	v.	经历，经受（变化、不快的事等）
ore	/ɔː(r)/	n.	矿；矿石；矿砂
smelt	/smelt/	v.	熔炼，提炼（金属）
bloomery	/ˈbluːməri/	n.	土法（木炭）熟铁吹炼炉
hammer	/ˈhæmə(r)/	v.	（用锤子）敲，锤打
furnace	/ˈfɜːnɪs/	n.	熔炉

43

toothpaste	/ˈtuːθpeɪst/	n.	牙膏
gum	/ɡʌm/	n.	牙龈；齿龈
pumice	/ˈpʌmɪs/	n.	浮岩
abrasiveness	/əˈbreɪsɪvnəs/	n.	耐磨性
powdered	/ˈpaʊdəd/	adj.	研成粉末的
charcoal	/ˈtʃɑːkəʊl/	n.	炭
ginseng	/ˈdʒɪnseŋ/	n.	人参；西洋参
chalk	/tʃɔːk/	n.	白垩
collapsible	/kəˈlæpsəbl/	adj.	可折叠的；可套缩的
tube	/tjuːb/	n.	管；管子
fluoride	/ˈflʊəraɪd/	n.	氟化物
cavity	/ˈkævəti/	n.	（龋齿的）窝洞，洞
oyster shell			牡蛎壳
herbal mint			香草薄荷
betel nut			槟榔果

44

indicator	/ˈɪndɪkeɪtə(r)/	n.	指针
weighing scale			秤；天平

45

ploughshare	/ˈplaʊʃeə(r)/	n.	铧；犁铧
plough	/plaʊ/	n./v.	犁

Words and Expressions
（生词与短语）

mouldboard	/ˈməʊldbɔːd/	n.	犁壁
coulter	/ˈkəʊltə/	n.	犁刀；犁头铁；铲
spike	/spaɪk/	n.	尖状物；尖头；尖刺
integral	/ɪnˈtegrəl/	adj.	完整的
blacksmith	/ˈblæksmɪθ/	n.	铁匠（尤指打马蹄铁者）

46

dentistry	/ˈdentɪstri/	n.	牙医业；牙科
decay	/dɪˈkeɪ/	v.	腐烂；腐朽
reside	/rɪˈzaɪd/	v.	居住在；定居于
dental	/ˈdentl/	adj.	牙齿的；牙科的
extract	/ɪkˈstrækt/	v.	（用力）取出，拔出
forceps	/ˈfɔːseps/	n.	（医生用的）镊子
porcelain	/ˈpɔːsəlɪn/	n.	瓷；瓷器

- -

bow drill			弓钻

47

excavate	/ˈekskəveɪt/	v.	挖掘；发掘
grave	/ɡreɪv/	n.	坟墓；墓地；墓穴
cemetery	/ˈsemətri/	n.	公墓
correspond	/ˌkɒrɪˈspɒnd/	v.	与……一致；相当于；符合
pastoralism	/ˈpɑːstərəlɪzəm/	n.	游牧（牧者带着牲口逐水草而居）
nomad	/ˈnəʊmæd/	n.	牧民
herd	/hɜːd/	n.	牧群
tunic	/ˈtjuːnɪk/	n.	束腰上衣
robe	/rəʊb/	n.	长袍
bumpy	/ˈbʌmpi/	adj.	不平的；多凸块的；颠簸的

48

pharaoh	/ˈfeərəʊ/	n.	法老（古埃及国王）
solely	/ˈsəʊlli/	adv.	只；仅；单独地
gravel	/ˈɡrævl/	n.	砾石；沙砾；石子
binder	/ˈbaɪndə(r)/	n.	黏合剂；黏结剂

- -

paved road			铺面道路
in a true sense			在真正意义上

49

awkward	/ˈɔːkwəd/	adj.	令人尴尬的；使人难堪的
lavatory	/ˈlævətri/	n.	抽水马桶；厕所；盥洗室
flush	/flʌʃ/	v.	冲洗；冲（抽水马桶）
luxury	/ˈlʌkʃəri/	n.	奢侈的享受；奢华
pipe	/paɪp/	n.	管子；管道
drain	/dreɪn/	v.	（使）排空；（使）流出

sophisticated	/sə'fɪstɪkeɪtɪd/	adj.	复杂巧妙的
flourish	/'flʌrɪʃ/	v.	繁荣；兴旺；昌盛；茁壮成长
latrine	/lə'tri:n/	n.	（营地等的）厕所；便坑
sewer	/'su:ə(r)/	n.	下水道；污水管道
answer nature's call			上厕所

50

concrete	/'kɒŋkri:t/	n.	混凝土；水泥
condensed	/kən'denst/	adj.	浓缩的
compact	/kəm'pækt/	adj.	紧密的；坚实的
rod	/rɒd/	n.	杆；竿；棒
aqueduct	/'ækwɪdʌkt/	n.	渡槽；高架渠
approximately	/ə'prɒksɪmətli/	adv.	大约；大概；约莫
quicklime	/'kwɪklaɪm/	n.	生石灰
pozzolana	/ˌpɒtsə'lɑ:nə/	n.	火山灰泥
aggregate	/'ægrɪgət/	n.	聚集体；集料
coarse	/kɔ:rs/	adj.	粗糙的；粗织的；粗的；大颗粒的
pebble	/'pebl/	n.	鹅卵石；砾石
finely	/'faɪnli/	adv.	成颗粒；细微地；细小地
limekiln	/laɪmkɪln/	n.	石灰窑
evaporate	/ɪ'væpəreɪt/	v.	蒸发，挥发
steel mesh			钢丝网

51

plier	/'plaɪə/	n.	钳子
necessity	/nə'sesəti/	n.	必需品；必然性
tong	/tɒŋ/	n.	钳子
blazing	/'bleɪzɪŋ/	adj.	熊熊燃烧的
makeshift	/'meɪkʃɪft/	adj.	临时替代的，权宜的
circa	/'sɜ:kə/	adv.	大概；大约；近于；近似
surgical	/'sɜ:(r)dʒɪk(ə)l/	adj.	外科的；外科手术的
laboratory	/lə'bɒrətri/	n.	实验室

52

resist	/rɪ'zɪst/	v.	抵制；抵抗；抵挡
trickle	/'trɪkl/	v.	滴，淌，小股流淌

53

realisation	/ˌri:əlaɪ'zeɪʃn/	n.	认识；领会；领悟
sacrifice	/'sækrɪfaɪs/	n.	祭祀；祭品
shower gel			沐浴露
come into existence			产生；出现

Words and Expressions
（生词与短语）

54
Minoan	/mɪ'nəʊən/	adj.	（以克里特岛为中心发展起来的）米诺斯文化的
Harappan	/hərə'pən/	adj.	（位于巴基斯坦的）哈拉帕文化的
install	/ɪn'stɔːl/	v.	安装；设置

55
source	/sɔːs/	n.	来源；源头；根源；原因
transfer	/træns'fɜː(r)/	v.	转移
indigo	/'ɪndɪɡəʊ/	n.	靛蓝；靛蓝类染料
sought-after	/'sɔːt ɑːftə(r)/	adj.	吃香的；广受欢迎的

56
column	/'kɒləm/	n.	柱；（通常为）圆形石柱；纪念柱
pillar	/'pɪlə(r)/	n.	（尤指兼作装饰的）柱子
distribute	/dɪ'strɪbjuːt/	v.	使散开；使分布；分散
ceiling	/'siːlɪŋ/	n.	天花板；顶棚
Assyrian	/ə'sɪriən/	adj.	与古亚述有关的
ornate	/ɔː'neɪt/	adj.	华美的；富丽的；豪华的
diminish	/dɪ'mɪnɪʃ/	v.	减少

57
paste	/peɪst/	n.	面团
lavish	/'lævɪʃ/	adj.	大量的；给人印象深刻的；耗资巨大的
spread	/spred/	n.	丰盛的餐食
innovative	/'ɪnəveɪtɪv/	adj.	采用新方法的；创新的
consume	/kən'sjuːm/	v.	吃；喝；饮
pastry	/'peɪstri/	n.	油酥面团；油酥面皮；油酥糕点
prestige	/pre'stiːʒ/	n.	声望；威望；威信
whereas	/weər'æz/	conj.	尽管；（用于正式文件中句子的开头）鉴于
folk	/fəʊk/	n.(pl.)	人们；普通百姓
settle for			无奈接受；勉强同意

58
repulse	/rɪ'pʌls/	v.	拒绝接受；回绝
cuisine	/kwɪ'ziːn/	n.	烹饪；菜肴；风味
settle in			定居于

59
plantation	/plɑːn'teɪʃn/	n.	种植园，种植场
hoe	/həʊ/	n.	锄头
frown	/fraʊn/	v.	皱眉；蹙眉头
be frowned upon			为人所唾弃
tie ... to ...			把……系到……上

60

fervour	/ˈfɜːvə(r)/	n.	热情；热诚；热烈
Latin	/ˈlætɪn/	adj.	拉丁语系国家（或民族）的
distill	/dɪsˈtɪl/	v.	提取……的精华
calamus	/ˈkæləməs/	n.	（植）菖蒲；省藤属
filter	/ˈfɪltə(r)/	v.	过滤
aromatic	/ˌærəˈmætɪk/	n.	香料；芳香植物
extract	/ɪkˈstrækt/	n.	提取物；浓缩物；精；汁
almond	/ˈɑːmənd/	n.	杏仁；扁桃仁
bergamot	/ˈbɜːɡəmɒt/	n.	香柠檬精油；香柠檬香草；香柠檬草药
myrtle	/ˈmɜːtl/	n.	香桃木；爱神木；番樱桃
set aside			搁置

61

arch	/ɑːtʃ/	n.	拱门
drainage	/ˈdreɪnɪdʒ/	n.	排水；排水系统
extensively	/ɪkˈstensɪvli/	adv.	广泛地；大量地
improvisation	/ˌɪmprəvaɪˈzeɪʃn/	n.	临时凑合的东西，即兴作品
triumphal	/traɪˈʌmfl/	adj.	庆祝成功（或胜利）的；凯旋的
monument	/ˈmɒnjumənt/	n.	纪念碑
true arch			纯圆形拱门
the triumphal arch			凯旋门

62

initially	/ɪˈnɪʃəli/	adv.	最初；开始；起初
cubit	/ˈkjuːbɪt/	n.	肘尺，腕尺（古代长度单位）
elbow	/ˈelbəʊ/	n.	肘；肘部
drawback	/ˈdrɔːbæk/	n.	缺点；不利条件

63

parasol	/ˈpærəsɒl/	n.	大遮阳伞
rank	/ræŋk/	n.	地位，级别
royalty	/ˈrɔɪəlti/	n.	王室成员
nobility	/nəʊˈbɪləti/	n.	贵族
fair skin			白皙的皮肤

64

chariot	/ˈtʃæriət/	n.	（古代用于战斗或比赛的）双轮敞篷马车
prominent	/ˈprɒmɪnənt/	adj.	重要的；著名的；杰出的；显眼的
spoke	/spəʊk/	n.	（车轮的）辐条；轮辐
Eurasian	/juˈreɪʒn/	adj.	欧亚的
peak	/piːk/	n.	高峰；顶峰；山峰；尖端
infantry	/ˈɪnfəntri/	n.	（统称）步兵

Words and Expressions
（生词与短语）

| charioteers | /ˌtʃæriəˈtɪə(r)/ | n. | 驾双轮马车的人 |
| command-vehicles | | | 指挥车 |

65
| spring scissors | | | 弹簧剪 |

66
sugarcane	/ˈʃʊɡəˌkeɪn/	n.	（植）甘蔗
particle	/ˈpɑːtɪkl/	n.	颗粒；微粒
tonne	/tʌn/	n.	公吨
steadily	/ˈstedəli/	adv.	稳定地；坚定地；不断地
refined sugar			精制糖

67
archaeological	/ˌɑːkiəˈlɒdʒɪkl/	adj.	考古学（上）的
artillery	/ɑːˈtɪləri/	n.	（统称）火炮
quench-hardened	/kwentʃ ˈhɑːdnd/	adj.	淬火硬化的
wrought	/rɔːt/	adj.	（金属）锻造的
intermediate	/ˌɪntəˈmiːdiət/	adj.	中级的；中等的；适合中等程度者的
speculate	/ˈspekjuleɪt/	v.	推测；猜测；推断

68
Archimedes' screw	/ˌɑːkɪˈmiːdiːz skruː/	n.	阿基米德式螺旋泵
shallow	/ˈʃæləʊ/	adj.	浅的
drain out			排出；流出

69
sudarium	/sjuːˈdeəriəm/	n.	手帕
wipe	/waɪp/	v.	（用布、手等）擦干净
pashmina	/pæʃˈmiːnə/	n.	山羊绒
warrior	/ˈwɒriə(r)/	n.	勇士
identify	/aɪˈdentɪfaɪ/	v.	显示；说明身份；确认；识别
plain	/pleɪn/	adj.	极普通的；朴素的
sweat cloth			汗巾

70
resemble	/rɪˈzembl/	v.	类似；相似；相像
elite	/ɪˈliːt/	n.	精英；杰出人物
commoner	/ˈkɒmənə(r)/	n.	平民
improvise	/ˈɪmprəvaɪz/	v.	临时拼凑；临时做；即兴创作
lead	/liːd/	n.	铅
intricate	/ˈɪntrɪkət/	adj.	错综复杂的
sanitation	/ˌsænɪˈteɪʃn/	n.	卫生设备；卫生设施体系
locker room			更衣室
cater to			迎合

71

incriminate	/ɪnˈkrɪmɪneɪt/	v.	使负罪；连累
gauntlet	/ˈɡɔːntlət/	n.	（中世纪武士铠甲的）金属手套
bribe	/braɪb/	n.	贿赂
occasional	/əˈkeɪʒənl/	adj.	偶尔的；偶然的
shorthand	/ˈʃɔːthænd/	n.	速记
ornament	/ˈɔːnəmənt/	n.	装饰品
linen	/ˈlɪnɪn/	n.	亚麻布

72

plumbing	/ˈplʌmɪŋ/	n.	（建筑物的）管路系统，自来水管道
pedestal	/ˈpedɪstl/	n.	基座
marble	/ˈmɑːbl/	n.	大理石

73

device	/dɪˈvaɪs/	n.	装置；仪器；器具；设备
pin	/pɪn/	n.	大头针

74

calliper	/ˈkælɪpə(r)/	n.	测径规；卡尺（用于测量管子、圆形物体的直径）
compass	/ˈkʌmpəs/	n.	圆规；两脚规
minute	/maɪˈnjuːt/	adj.	极小的；微小的；细微的
specimen	/ˈspesɪmən/	n.	样品；样本；标本
shipwreck	/ˈʃɪprek/	n.	失事的船；沉船
inscription	/ɪnˈskrɪpʃn/	n.	（石头或金属上）刻写的文字，铭刻，碑文

75

terracotta	/ˌterəˈkɒtə/	n.	（无釉的）赤陶土，赤陶
downward	/ˈdaʊnwəd/	adj.	下降的；向下的
slope	/sləʊp/	n.	斜坡；坡度；山坡
gravity	/ˈɡrævəti/	n.	重力；地球引力
maintenance	/ˈmeɪntənəns/	n.	维护；维持；保养
thriving	/ˈθraɪvɪŋ/	adj.	欣欣向荣的，兴旺发达的
a stretch of			一片；一泓；一段（土地或水域）
fall out of			停止；放弃

76

catapult	/ˈkætəpʌlt/	n.	（旧时的）石弩，弩炮
fling	/flɪŋ/	v.	扔，掷，抛，丢
tyrant	/ˈtaɪrənt/	n.	暴君；专制君主
ballista	/bəˈlɪstə/	n.	（古代战争中的）弩炮
load	/ləʊd/	n.	负荷；负载；大量

Words and Expressions
（生词与短语）

77

barrel	/'bærəl/	n.	桶
revolutionary	/ˌrevə'luːʃənəri/	adj.	革命的；革命性的；巨变的
ridiculous	/rɪ'dɪkjuləs/	adj.	愚蠢的；荒谬的
fragile	/'frædʒaɪl/	adj.	易碎的；易损的；不牢固的
fro	/frəʊ/	adv.	来回地
stack	/stæk/	v.	（使）放成整齐的一叠（或一摞、一堆）
in comparison to			与……相比
come into the picture			出现
to and fro			来回地

78

magnifying glass	/'mæɡnɪfaɪɪŋ ɡlɑːs/	n.	放大镜
playwright	/'pleɪraɪt/	n.	剧作家；编剧
start fires			生火
glass globe			玻璃罩子

79

wheelbarrow	/'wiːlbærəʊ/	n.	独轮手推车
supply	/sə'plaɪ/	n.	供应；供给；提供；补给
ensure	/ɪn'ʃʊə(r)/	v.	保证；担保；确保

80

pulley	/'pʊli/	n.	滑轮；滑轮组
hoist	/hɔɪst/	v.	提升；吊起；拉高
compound	/'kɒmpaʊnd/	adj.	复合的
haul	/hɔːl/	v.	（用力）拖，拉，拽
considerable	/kən'sɪdərəbl/	adj.	相当多（或大、重要等）的
entire	/ɪn'taɪə(r)/	adj.	全部的；整个的；完全的
remarkable	/rɪ'mɑːkəbl/	adj.	非凡的；奇异的；显著的；引人注目的
stem from			出于；基于

81

manually	/'mænjuəli/	adv.	手工地；用手地
current	/'kʌrənt/	n.	（海洋或江河的）水流
groove	/ɡruːv/	n.	槽；沟；辙；纹
kinetic	/kaɪ'netɪk/	adj.	运动的；运动引起的
deposit	/dɪ'pɒzɪt/	v.	放下；放置
churn	/tʃɜːn/	v.	剧烈搅动；（使）猛烈翻腾
mill	/mɪl/	n.	磨坊；面粉厂；工厂
simultaneously	/ˌsɪml'teɪniəsli/	adv.	同时发生（或进行）地；同步地
turbine	/'tɜːbaɪn/	n.	汽轮机；涡轮机
rotor	/'rəʊtə(r)/	n.	（机器的）转子

motion	/ˈməʊʃn/	n.	运动；移动
power generation			发电

82

slightly	/ˈslaɪtli/	adv.	稍微；略微
taper	/ˈteɪpə(r)/	v.	（使）逐渐变窄
diameter	/daɪˈæmɪtə(r)/	n.	直径；对径
faucet	/ˈfɔːsɪt/	n.	龙头；旋塞
jet	/dʒet/	v.	射出；喷出
conceal	/kənˈsiːl/	v.	隐藏；掩盖
water closet	/ˈwɔːtə(r) ˈklɒzɪt/	n.	厕所；抽水马桶
reservoir	/ˈrezəvwɑː(r)/	n.	水库；蓄水池
empty into			注入；流注；倾入

83

compass	/ˈkʌmpəs/	n.	罗盘；指南针
mineral	/ˈmɪnərəl/	n.	矿物；矿物质
lodestone	/ˈləʊdˌstəʊn/	n.	磁铁矿；天然磁石
compose	/kəmˈpəʊz/	v.	组成，构成（一个整体）
orient	/ˈɔːrɪənt/	v.	使朝向；使面对；确定方向
constellation	/ˌkɒnstəˈleɪʃn/	n.	星座
magnetised	/ˈmæɡnəˌtaɪzd/	adj.	磁化的
pivot	/ˈpɪvət/	n.	支点；枢轴；中心点
magnetometer	/ˌmæɡnɪˈtɒmɪtə/	n.	磁强计；地磁仪
embed	/ɪmˈbed/	v.	把……牢牢地嵌入（或插入、埋入）
in use			正在用
be composed of			由……组成
fortune teller			算命卖卜者

84

palanquin	/ˌpælənˈkiːn/	n.	（东方国家旧时用人抬的）四或六人大轿
porter	/ˈpɔːtə(r)/	n.	脚夫
litter	/ˈlɪtə(r)/	n.	（旧时抬要人的）轿，舆
backpack	/ˈbækpæk/	n.	（箱型）背包
landscape	/ˈlændskeɪp/	n.	山水画；乡村风景画
mandarin	/ˈmændərɪn/	n.	政界要员；（尤指）内务官员
enclose	/ɪnˈkləʊz/	v.	围住
bridal	/ˈbraɪdl/	adj.	新娘的；婚礼的
utmost	/ˈʌtməʊst/	adj.	最大的；极度的
deck	/dek/	v.	装饰；布置；打扮
auspicious	/ɔːˈspɪʃəs/	adj.	吉利的；吉祥的
ornament	/ˈɔːnəmənt/	v.	装饰；点缀；美化

Words and Expressions
（生词与短语）

gild	/ˈɡɪld/	v.	给……镀金
screen	/skriːn/	v.	遮蔽
be equipped with			配有

85

candlestick	/ˈkændlstɪk/	n.	蜡烛台（或架）
prong	/prɒŋ/	n.	叉子齿；叉齿
wax	/wæks/	n.	蜡；石蜡；蜂蜡
cinnamon	/ˈsɪnəmən/	n.	肉桂皮
eulachon	/ˈjuːləkɒn/	n.	太平洋细齿鲑［亦称candlefish］
relatively	/ˈrelətɪvli/	adv.	相当程度上；相当地；相对地
tell time			报时

86

seesaw	/ˈsiːsɔː/	n.	跷跷板
beam	/biːm/	n.	梁；平衡木
fulcrum	/ˈfʊlkrəm/	n.	（杠杆的）支点，支轴
gauge	/ɡeɪdʒ/	v.	估计；判定；估算
force of gravity			地心引力

87

dome	/dəʊm/	n.	穹顶；圆顶状物；穹状建筑物
tusk	/tʌsk/	n.	（象和某些其他动物的）长牙
igloo	/ˈɪɡluː/	n.	（北美北部因纽特人的拱形圆顶）冰屋
pioneer	/ˌpaɪəˈnɪə(r)/	v.	当开拓者；做先锋
legacy	/ˈleɡəsi/	n.	传统；传承
hold up			举起；支撑
carry forward			推进；发扬；使进行

88

milling	/ˈmɪlɪŋ/	n.	磨；制粉；碾碎
grind	/ɡraɪnd/	v.	研磨；磨碎［过去式和过去分词是ground］
fine	/faɪn/	adj.	小颗粒制成的；颗粒细微的
separate	/ˈsepəreɪt/	v.	分离；隔开
reclamation	/ˌrekləˈmeɪʃn/	n.	开垦
mortar	/ˈmɔːtə(r)/	n.	臼；研钵
pestle	/ˈpesl/	n.	杵、碾槌
land fill			土地填筑

89

lunisolar	/ˌluːnɪˈsəʊlə/	adj.	日月的；阴阳的
longitude	/ˈlɒŋɡɪtjuːd; ˈlɒndʒɪtjuːd/	n.	经度
phase	/feɪz/	n.	月相；（月亮的）盈亏

course	/kɔ:s/	n.	轨迹
civil	/'sɪvl/	adj.	国民的；平民的
determine	/dɪ'tɜ:mɪn/	v.	决定；测定；准确算出
in sync with			与……一致；与……同步
leap month			（天）闰月
be traced back to			追溯到……

90

thermally	/'θɜ:məli/	adv.	热地；用热的方法
insulated	/'ɪnsjuleɪtɪd/	adj.	有隔热（或隔音、绝缘）保护的
chamber	/'tʃeɪmbə(r)/	n.	（机器内的）腔，室
pit	/pɪt/	n.	深洞；深坑
tile	/taɪl/	n.	瓦，瓦片
stove	/stəʊv/	n.	炉
stew	/stju:/	n.	炖煮的菜肴
pre-dynastic	/ˌpriːdɪ'næstɪk/	adj.	（尤指公元前3000年左右的古埃及）与王朝统治以前有关的
settlement	/'setlmənt/	n.	定居点；居民点

91

origin	/'ɒrɪdʒɪn/	n.	起源；起因；源头
relay	/ri:'leɪ/	n.	接班的人（或动物）；轮换者
convey	/kən'veɪ/	v.	输送；运送；传送
courier	/'kʊriə(r)/	n.	（递送包裹或重要文件的）信使
encounter	/ɪn'kaʊntə(r)/	v.	偶然碰到
curcus publicus			（罗马帝国的）公共邮政

92

witness	/'wɪtnəs/	v.	见证
recreation	/ˌrekri'eɪʃn/	n.	娱乐；消遣；娱乐活动；游戏
bloomers	/'blu:məz/	n.	女士灯笼裤
drawers	/'drɔ:z/	n.	（旧或幽默）衬裤，内裤
exposure	/ɪk'spəʊʒə(r)/	n.	暴露；显露；露出
waist	/weɪst/	n.	腰；（衣服的）腰部
ruffled	/'rʌfld/	adj.	有褶饰边的
attire	/ə'taɪə(r)/	n.	服装；衣服
straw hat			草帽

93

matted	/'mætɪd/	n.	缠结的；湿脏蓬乱的
woven	/'wəʊvən/	adj.	编织的
sew	/səʊ/	v.	缝；缝纫；缝合；缝制
knit	/nɪt/	v.	编织；针织；机织；织平针

Words and Expressions
（生词与短语）

leggings	/ˈleɡɪŋz/	n.	绑腿；护腿；女式紧身裤；裹腿
nylon	/ˈnaɪlɒn/	n.	尼龙；尼龙连袜裤
blend	/blend/	v.	混合；融合
currently	/ˈkʌrəntli/	adv.	当前；目前；现时；时下
anklets	/ˈæŋkləts/	n.	翻口短袜
bare	/beə(r)/	adj.	裸体的；裸露的
toe socks			五趾袜
bare socks			隐形袜

94

moist	/mɔɪst/	adj.	微湿的；湿润的
rag	/ræɡ/	n.	抹布
pith	/pɪθ/	n.	（橙子等水果中的）果皮
shrub	/ʃrʌb/	n.	灌木
imperial court			朝廷
cellulose pulp			纤维素纸浆
mulberry bark			桑树树皮
rice straw			稻草

95

vertical	/ˈvɜːtɪkl/	adj.	竖的；垂直的；直立的；纵向的

96

frayed	/freɪd/	adj.	磨损的
stiff	/stɪf/	adj.	不易弯曲的；硬的；挺的
bristle	/ˈbrɪsl/	n.	刷子毛

97

vault	/vɔːlt/	n.	拱顶；穹顶
aspect	/ˈæspekt/	n.	外观
tunnel/barrel vault	/ˈtʌnəl vɔːlt; ˈbærəl vɔːlt/	n.	筒形穹窿
with regards to			关于

98

hydrometer	/haɪˈdrɒmɪtə(r)/	n.	液体比重计
credit	/ˈkredɪt/	n.	功劳；声望
density	/ˈdensəti/	n.	密度；浓度
cylindrical	/səˈlɪndrɪkl/	adj.	圆柱形的；圆筒状的
stem	/stem/	n.	柄
mercury	/ˈmɜːkjəri/	n.	汞；水银
cylinder	/ˈsɪlɪndə(r)/	n.	（尤指用作容器的）圆筒状物
scale	/skeɪl/	n.	比例尺；刻度
be given the credit for ...			因……被赞扬

99

realistic	/ˌriːəˈlɪstɪk/	adj.	现实的；实际的；实事求是的；明智的
astronomer	/əˈstrɒnəmər/	n.	天文学家
astrologer	/əˈstrɒlədʒə(r)/	n.	占星家
obsess	/əbˈses/	v.	使迷恋
horoscope	/ˈhɒrəskəʊp/	n.	占星术
precisely	/prɪˈsaɪsli/	adv.	精确地；准确地；恰好地；细心地
augment	/ɔːɡˈment/	v.	增加；增强；扩大；充填
latitude	/ˈlætɪtjuːd/	n.	纬度；纬度地区
plot	/plɒt/	v.	（在地图上）画出；绘制（图表）
flatten	/ˈflætn/	v.	（使）变平；把……弄平；摧毁；轻易击败
dimensional	/dɪˈmenʃənəl/	adj.	空间的；（数）因次的；……次（元）的
Babylonian	/ˌbæbɪˈləʊnɪən/	adj.	巴比伦的
tablet	/ˈtæblət/	n.	牌，碑，匾
resemblance	/rɪˈzembləns/	n.	相似；相像
depict	/dɪˈpɪkt/	v.	描画；描写；叙述
circular	/ˈsɜːkjələ(r)/	adj.	圆形的；环形的；圆的；环行的
bisect	/baɪˈsekt/	v.	对开；平分；两分；（数）二等分
navigation	/ˌnævɪˈɡeɪʃn/	n.	导航；航行；领航
extensive	/ɪkˈstensɪv/	adj.	广阔的；广大的；大量的；广泛的
dawn	/dɔːn/	v.	（一天或一个时期）开始
voyage	/ˈvɔɪɪdʒ/	n.	航行；（尤指）航海

100

cumbersome	/ˈkʌmbəsəm/	adj.	大而笨重的；难以携带的
metallic	/məˈtælɪk/	adj.	金属制的
transaction	/trænˈzækʃn/	n.	交易；处理；业务
deposit	/dɪˈpɒzɪt/	n./v.	存款；存储
individual	/ˌɪndɪˈvɪdʒuəl/	n.	个人
trustworthy	/ˈtrʌstˌwɜːði/	adj.	值得信任的；可信赖的；可靠的
denote	/dɪˈnəʊt/	v.	表示
redeem	/rɪˈdiːm/	v.	兑换；兑现

101

cannon	/ˈkænən/	n.	（通常装有轮子并发射铁弹或石弹的旧式）大炮
truncated	/trʌŋˈkeɪtɪd/	adj.	截短的；截平的；截成平面的
cone	/kəʊn/	n.	（实心或空心的）圆锥体；
internal	/ɪnˈtɜːnl/	adj.	内部的；里面的
bore	/bɔː(r)/	n.	（管道、枪炮等的）内径；膛径
projectile	/prəˈdʒektaɪl/	n.	射弹；抛射体；火箭
explosive charge			炸药装填量

Words and Expressions
（生词与短语）

102

pretzel	/'pretsl/	n.	椒盐卷饼（常作小吃）
monk	/mʌŋk/	n.	僧侣；修道士
treat	/triːt/	n.	款待
motivate	/'məʊtɪveɪt/	v.	激励；激发
distracted	/dɪ'stræktɪd/	adj.	注意力分散的；思想不集中的
emblem	/'embləm/	n.	象征；标志；标记
seasoning	/'siːzənɪŋ/	n.	调味品；佐料
complement	/'kɒmplɪment/	v.	补充；补足
washing soda	/'wɒʃɪŋ 'səʊdə/	n.	洗用碱
lye	/laɪ/	n.	碱液
flavour	/'fleɪvə(r)/	n.	（食物或饮料的）味道；（某种）味道
glaze	/gleɪz/	n.	（浇在糕点上增加光泽的）奶浆，糖浆
in prayer			在祈祷

103

prototype	/'prəʊtətaɪp/	n.	原型；雏形；最初形态
normally	/'nɔːməli/	adv.	通常；正常地
tubular	/'tjuːbjʊlə(r)/	adj.	管子构成的；有管状部分的
discharge	/dɪs'tʃɑːdʒ/	v.	射出；开火
fire lance			火枪
projectile weapon			投掷型武器
range from			从……到；范围从……到

104

alchemist	/'ælkəmɪst/	n.	炼金术士；炼金师
immortality	/ˌɪmɔː'tæləti/	n.	不朽；永生；不灭
explode	/ɪk'spləʊd/	v.	爆炸；爆裂；爆破
scare	/skeə(r)/	v.	惊吓；使害怕
explosive	/ɪk'spləʊsɪv/	n.	炸药；爆炸物
Arab	/'ærəb/	adj.	阿拉伯的

105

velvet	/'velvɪt/	n.	天鹅绒；丝绒
dense	/dens/	adj.	密集的；稠密的
unique	/juː'niːk/	adj.	独一无二的；独特的
synthetic	/sɪn'θetɪk/	adj.	人造的；（人工）合成的
loom	/luːm/	n.	织布机
technique	/tek'niːk/	n.	技术；方法；手法
fairly	/'feəli/	adv.	相当地
manufacture	/ˌmænju'fæktʃə(r)/	v.	生产
extremely	/ɪk'striːmli/	adv.	非常；极其

106

emerge	/ɪ'mɜːdʒ/	v.	浮现；出现
professional	/prə'feʃənl/	adj.	职业的；专业的
pioneer	/ˌpaɪə'nɪə(r)/	n.	先锋；先驱；带头人
principle	/'prɪnsəpl/	n.	原理；原则
curiosity	/ˌkjʊəri'ɒsəti/	n.	好奇心；求知欲
rocketry	/'rɒkɪtri/	n.	火箭学；火箭技术
launch	/lɔːntʃ/	v.	（航天器的）发射
liquid-propellant	/'lɪkwɪd prə'pelənt/	n.	液体推进剂
due to			由于

107

credit	/'kredɪt/	v.	认为是……的功劳；把……归于
overlay	/ˈəʊvəˌleɪ/	v.	覆盖；包
showcase	/'ʃəʊkeɪs/	v.	展示
status	/'steɪtəs/	n.	地位；身份
gradually	/'grædʒuəli/	adv.	渐渐地；逐步地
peasant	/'peznt/	n.	农民
gown	/gaʊn/	n.	（尤指特别场合穿的）女裙；女长服；女礼服
strata	/'strɑːtə/	n.	阶层［单数是stratum］
right up to			直到
be reduced to			沦为；沦落为

108

spinning wheel	/'spɪnɪŋ wiːl/	n.	纺车
spindle	/'spɪndl/	n.	（手纺用的）绕线标；纺锤
card	/kɑːd/	v.	（用钢丝刷）梳理
		n.	梳理机
loosely	/'luːsli/	adv.	松散地；宽松地；松弛地
yarn	/jɑːn/	n.	纱；纱线
fleecy	/'fliːsi/	adj.	软如羊毛的；羊毛似的
candlewick	/'kændlwɪk/	n.	烛芯纱（有凸起花纹，尤用于制作床罩）
reel	/riːl/	n.	卷筒；卷轴；卷盘
bobbin	/'bɒbɪn/	n.	线轴；绕线筒
attach to			连接在……上；附于
comb with			梳理
spin into			旋入；卷入

109

firearm	/'faɪərɑːm/	n.	（便携式的）枪
gunsmith	/'gʌnsmɪθ/	n.	造枪工；修枪匠；军械工人
flintlock	/'flɪntlɒk/	n.	（旧时的）燧发机；明火枪

Words and Expressions
（生词与短语）

helical groove	/ˈhelɪkl ɡruːv/	n.	螺旋槽
trigger	/ˈtrɪɡə(r)/	n.	（枪的）扳机
spring-loaded	/sprɪŋ ˈləʊdɪd/	adj.	弹簧承载的；弹顶的
flint	/flɪnt/	n.	燧石；火石
ignite	/ɪɡˈnaɪt/	v.	点燃；（使）燃烧
propel	/prəˈpel/	v.	推进；推动；驱动
spherical	/ˈsferɪkl/	adj.	球形的；球状的
bullet	/ˈbʊlɪt/	n.	子弹；弹丸
formerly	/ˈfɔːməli/	adv.	以前；从前
squeeze	/skwiːz/	n.	挤压；捏
round	/raʊnd/	n.	一次射击；一发子弹
burst	/bɜːst/	n.	一阵短促的射击

110

delicate	/ˈdelɪkət/	adj.	精致的
openwork	/ˈəʊpənwɜːk/	adj.	透空式的
fancy	/ˈfænsi/	adj.	精致的；有精美装饰的
previously	/ˈpriːviəsli/	adv.	预先；在前
decor	/ˈdeɪkɔː(r)/	n.	（建筑内部的）装饰布局，装饰风格

give the credit to			归功于
gain a lot of popularity			大受欢迎
make an appearance			出场；出现
home decor			家居装饰

111

spectacles	/ˈspektəklz/	n.	（正式）眼镜
convex	/ˈkɒnveks/	n.	凸面
crystal	/krɪstl/	n.	水晶；结晶体
rivet	/ˈrɪvɪt/	n.	铆钉
magnifier	/ˈmæɡnɪfaɪə(r)/	n.	放大镜；放大器
lay	/leɪ/	adj.	外行的；非专业的；缺少专门知识的
optician	/ɒpˈtɪʃn/	n.	眼镜商；验光师
rigid	/ˈrɪdʒɪd/	adj.	坚硬的；不弯曲的
hook	/hʊk/	v.	（使）钩住，挂住
function	/ˈfʌŋkʃn/	n.	功能；作用
magnification	/ˌmæɡnɪfɪˈkeɪʃn/	n.	放大；放大倍数

| make a profit | | | 获利；赚钱；赚取利润 |
| prescription sunglasses | | | 有度数的墨镜 |

112

| screwdriver | /ˈskruːdraɪvə(r)/ | n. | 螺丝刀；改锥 |
| cork | /kɔːk/ | n. | （尤指酒瓶的）软木塞 |

olive	/ˈɒlɪv/	n.	橄榄；油橄榄
durability	/ˌdjʊərəˈbɪləti/	n.	耐用性；稳定性；耐用度
stability	/stəˈbɪləti/	n.	稳定（性）；稳固（性）
accompany	/əˈkʌmp(ə)ni/	v.	伴随；与……同时发生
screw	/skruː/	n.	螺丝；螺丝钉
nut	/nʌt/	n.	螺母；螺帽
fasten	/ˈfɑːsn/	v.	使牢固；使固定
tighten	/ˈtaɪt(ə)n/	v.	（使）变紧，更加牢固
bolt	/bəʊlt/	n.	螺栓
loosen	/ˈluːs(ə)n/	v.	松开；解开

113

amber	/ˈæmbə(r)/	n.	（矿）琥珀
static electricity	/ˈstætɪk ɪˌlekˈtrɪsəti/	n.	静电
physician	/fɪˈzɪʃn/	n.	内科医生
magnetism	/ˈmæɡnətɪzm/	n.	磁性；磁力
name ... after ...			以……命名……

114

deposit	/dɪˈpɒzɪt/	n.	（地下自然形成的）沉积物，沉积层
graphite	/ˈɡræfaɪt/	n.	石墨
resident	/ˈrezɪdənt/	n.	居民；住户
misidentify	/ˌmɪsaɪˈdentɪfaɪ/	v.	错误识别
brittle	/ˈbrɪtl/	adj.	硬但易碎的；脆性的
juniper	/ˈdʒuːnɪpə(r)/	n.	桧柏

115

locksmith	/ˈlɒksmɪθ/	n.	锁匠；修锁工
portable	/ˈpɔːtəbl/	adj.	便携式的；手提的；轻便的
wind	/waɪnd/	v.	给（钟表等）上发条
spring	/sprɪŋ/	n.	弹簧；发条
hand	/hænd/	n.	指针
rotate	/rəʊˈteɪt/	v.	旋转；转动
span	/spæn/	v.	持续；贯穿
keep account of			记录
be devoted to doing something			致力于做某事

116

evenly	/ˈiːvnli/	adv.	平均地；均等地
movable	/ˈmuːvəbl/	adj.	可移动的
character	/ˈkærəktə(r)/	n.	（书写、印刷或计算机上的）文字，字母，符号
smooth	/smuːð/	adj.	平整的，平坦的
refinement	/rɪˈfaɪnmənt/	n.	（精细的）改进，改善

Words and Expressions
（生词与短语）

mechanisation	/ˌmekənaɪˈzeɪʃən/	n.	机械化
principal	/ˈprɪnsəpl/	adj.	最重要的；主要的
promote	/prəˈməʊt/	v.	促进；推动
literacy	/ˈlɪtərəsi/	n.	读写能力；素养；文化水平
regard … as …			认为；把……认作；视为
promote literacy			扫盲

117

corset	/ˈkɔːsɪt/	n.	（尤指旧时妇女束腰的）紧身内衣
torso	/ˈtɔːsəʊ/	n.	（身体的）躯干
garment	/ˈɡɑːmənt/	n.	（一件）衣服
bodice	/ˈbɒdɪs/	n.	连衣裙上身
stay	/steɪ/	n.	（妇女的）紧身褡，胸衣
whalebone	/ˈweɪlbəʊn/	n.	鲸须，鲸骨（旧时用以支撑衣服）
buckram	/ˈbʌkrəm/	n.	（旧时用作书皮或衣服衬里的）硬棉布
vertically	/ˈvɜːtɪkli/	adv.	垂直地
prior to			前于

118

unglazed	/ˌʌnˈɡleɪzd/	adj.	未上光的；未上釉的
spout	/spaʊt/	n.	（茶壶等的）嘴
exclusively	/ɪkˈskluːsɪvli/	adv.	专门地；特定地
brew	/bruː/	v.	沏；沏（茶）
withstand	/wɪðˈstænd/	v.	承受；经受住；抵住
Dutch	/dʌtʃ/	adj.	荷兰的
importer	/ɪmˈpɔːtə(r)/	n.	从事进口的人（或公司等）；进口商
chest	/tʃest/	n.	（常为木制的）大箱子
heat-resistant	/hiːt rɪˈzɪstənt/	adj.	耐热的
translucent	/trænsˈluːsnt/	adj.	半透明的
exquisite	/ɪkˈskwɪzɪt/	adj.	精美的；精致的
authentic	/ɔːˈθentɪk/	adj.	真正的；真品的；真迹的
covet	/ˈkʌvət/	v.	垂涎；渴望；贪求
enthusiast	/ɪnˈθjuːziæst/	n.	爱好者；热衷于……的人；热心者
bone china			骨质瓷，骨质瓷器（用瓷土与骨灰混合烧制成）

119

lens	/lenz/	n.	透镜；镜片
novelty	/ˈnɒvlti/	n.	新颖；新奇的事物（或人、环境）
maximum	/ˈmæksɪməm/	adj.	最高的；最大极限的
blurry	/ˈblɜːri/	adj.	模糊的
minimum	/ˈmɪnɪməm/	n.	最低限度；最小值；最少量
objective	/əbˈdʒektɪv/	n.	（望远镜或显微镜的）物镜

eyepiece	/'aɪpiːs/	n.	（望远镜或显微镜的）目镜
polish	/'pɒlɪʃ/	v.	抛光；磨光
convex	/'kɒnveks/	adj.	凸出的；凸面的
prolific	/prə'lɪfɪk/	adj.	多产的；创作丰富的
to a certain extent			在一定程度上；在某种程度上；达到某种程度

120

Persian	/'pɜːʃn/	adj.	波斯的
stirrup	/'stɪrəp/	n.	马镫
aristocracy	/ˌærɪ'stɒkrəsi/	n.	（某些国家的）贵族
masculine	/'mæskjəlɪn/	adj.	男子汉的；男人的；像男人的
trend	/trend/	n.	趋势；动向；趋向
accredited	/ə'kredɪtɪd/	adj.	官方认可的；获正式承认的
petite	/pə'tiːt/	adj.	纤弱的；娇小的
catch on			流行；受欢迎
be adopted by			被……所采用

121

hosiery	/'həʊziəri/	n.	（尤用于商店）袜类
hose	/həʊz/	n.	长筒袜
slave	/sleɪv/	n.	奴隶
evolve into			发展成
come into fashion			流行起来

122

pointed	/'pɔɪntɪd/	adj.	尖的；有尖头的
shoot	/ʃuːt/	v.	射击［过去式和过去分词是shot］
ammunition	/ˌæmju'nɪʃn/	n.	弹药
cartridge	/'kɑːtrɪdʒ/	n.	弹夹
contain	/kən'teɪn/	v.	包含；含有
core	/kɔː(r)/	adj.	最重要的；主要的；基本的
insert ... into			把……插入；把……嵌入

123

affiliation	/əˌfɪli'eɪʃn/	n.	（与政党、宗教等的）隶属关系
starkly	/'stɑːkli/	adv.	完全地；明显地
milliner	/'mɪlɪnə(r)/	n.	女帽制造商；制造（或销售）女帽的人
political affiliation			政治立场；政治面貌

124

wrinkle	/'rɪŋkl/	n.	褶皱
hire	/'haɪə(r)/	v.	雇用；聘用
pleat	/pliːt/	n.	（布料上缝的）褶
smoother	/'smuːðə(r)/	n.	平整工具；弄平用具

Words and Expressions
（生词与短语）

| constitute | /'kɒnstɪtjuːt/ | v. | 组成；构成 |
| goffering iron | | | 烫皱褶熨斗 |

125

propose	/prə'pəʊz/	v.	建议；提议
error	/'erə(r)/	n.	错误；差错；谬误
declare	/dɪ'kleə(r)/	v.	宣布；宣告；宣称
approximate	/ə'prɒksɪmeɪt/	v.	近似计算；概略估算
shift	/ʃɪft/	v.	转移；改变
approximation	/əˌprɒksɪ'meɪʃn/	n.	近似值；粗略估算
with respect to			（正式）考虑到

126

thermometer	/θə'mɒmɪtə(r)/	n.	温度计；体温计
particular	/pə'tɪkjələ(r)/	adj.	不寻常的；格外的，特别的
expand	/ɪk'spænd/	v.	扩大；扩展；膨胀
contract	/'kɒntrækt/	v.	收缩
thermoscope	/θɜː'məˌskəʊp/	n.	验温器
clinical	/'klɪnɪk(ə)l/	adj.	临床的；临床诊断的
enclosed	/ɪn'kləʊzd/	adj.	封闭的
standardised	/'stændəˌdaɪzd/	adj.	标准化的
alcohol thermometer			酒精温度计
numerical scale			数值刻度；数字比例尺
in a ... manner			以一种……的方式

127

wagonway	/'wægənweɪ/	n.	带轨的畜力车道（铁轨出现之前）
comprise	/kəm'praɪz/	v.	组成，构成
draw	/drɔː/	v.	拖，拉［过去式是drew，过去分词是drawn］
cart	/kɑːt/	n.	（两轮或四轮）运货马车
replace	/rɪ'pleɪs/	v.	取代；代替
tramway	/'træmweɪ/	n.	有轨电车轨道
flanged	/flændʒd/	adj.	带凸缘的
flange	/flændʒ/	n.	凸缘；（火车的）轮缘
grip	/ɡrɪp/	n.	手柄，把手，紧握
beneficial	/ˌbenɪ'fɪʃl/	adj.	有利的；有裨益的；有用的
aid	/eɪd/	v.	帮助，协助
locomotive	/ˌləʊkə'məʊtɪv/	n.	机车；火车头

128

plug	/plʌɡ/	v.	堵塞，封堵
cork oak	/kɔːk əʊk/	n.	栓皮栎树
accumulate	/ə'kjuːmjəleɪt/	v.	积累，聚集

stopper	/'stɒpə/	n.	瓶塞
cottage	/'kɒtɪdʒ/	n.	小木屋，农舍
harvest	/'hɑːvɪst/	v.	收获，收割，得到
wildly	/'waɪldli/	adv.	极其，非常
cultivation	/ˌkʌltɪ'veɪʃn/	n.	栽培

129

bow tie	/ˌbəʊ'taɪ/	n.	蝶形领结
collar	/'kɒlə(r)/	n.	衣领，领子
tuxedo	/tʌk'siːdəʊ/	n.	（配蝶形领结的）成套无尾晚礼服
tailcoat	/'teɪlkəʊt/	n.	燕尾服，男子晚礼服
complement	/'kɒmplɪment/	n.	补充物；补足物

130

submarine	/ˌsʌbmə'riːn/	n.	潜水艇
sketch	/sketʃ/	n.	素描；草图
wrap	/ræp/	v.	包，用……包裹
rowboat	/'rəʊˌbəʊt/	n.	划艇
air tube	/eə(r) tjuːb/	n.	空气管，空气导管
float	/fləʊt/	n.	浮子，鱼漂
gasket	/'gæskɪt/	n.	垫圈；衬垫；密封垫
propeller	/prə'pelə(r)/	n.	螺旋桨
navy	/'neɪvi/	n.	海军，海军部队
purchase	/'pɜːtʃəs/	v.	购买，采购
nuclear-powered	/'njuːkliə(r) 'paʊəd/	adj.	核动力的
launch	/lɔːntʃ/	v.	使（船，尤指新船）下水

131

pump	/pʌmp/	n.	泵
piston	/'pɪstən/	n.	活塞
condenser	/kən'densə(r)/	n.	冷凝器
rotary	/'rəʊtəri/	adj.	转动的
loaded	/'ləʊdɪd/	adj.	装载的；满载而沉重的

132

telescope	/'telɪˌskəʊp/	n.	望远镜
contrary	/'kɒntrəri/	adj.	相对立的；相反的
claim	/kleɪm/	v.	声称；宣称
patent	/'pætnt/	n.	专利
astronomy	/ə'strɒnəmi/	n.	天文学
crater	/'kreɪtə(r)/	n.	火山口；弹坑；弧坑；环形山
sunspot	/'sʌnspɒt/	n.	（太阳）黑子
moon	/muːn/	n.	卫星

Words and Expressions
（生词与短语）

contrary to			跟……相反
apply for			寻求；申请
positive lens			正透镜
narrow tube			小直径管
negative lens			负透镜

133

rest	/rest/	v.	（被）支撑；（使）倚靠；托
knot	/nɒt/	v.	把……打成结（或扎牢）
debatable	/dɪ'beɪtəbl/	adj.	可争辩的；有争议的
Croatian	/krəʊ'eɪʃən/	adj.	克罗地亚人的
cravate	/krə'væt/	n.	（男用）阔领带
reserve	/rɪ'zɜː(r)v/	v.	保留
sport	/spɔː(r)t/	v.	得意地穿戴
set the trend			（在风尚、式样上）带个头；创立新款式

134

barometer	/bə'rɒmɪtə(r)/	n.	气压计；晴雨表
vacuum	/'vækjuəm/	n.	真空状态
untimely	/ʌn'taɪmli/	adj.	突然的
be derived from			由……而来
air pressure			气压

135

transfusion	/træns'fjuːʒn/	n.	输血
vein	/veɪn/	n.	静脉
plasma	/'plæzmə/	n.	血浆
clotting factor	/'klɒtɪŋ 'fæktə(r)/	n.	凝血因子
platelet	/'pleɪtlət/	n.	血小板
circulation	/ˌsɜːkjə'leɪʃn/	n.	血液循环
property	/'prɒpəti/	n.	性质；特性；性能
revive	/rɪ'vaɪv/	v.	复活；（使）苏醒
artery	/'ɑːtəri/	n.	动脉
transfuse	/træns'fjuːz/	v.	输血
labourer	/'leɪbərə(r)/	n.	工人；体力劳动者
medical practice			医疗实践

136

parachute	/'pærəʃuːt/	n.	降落伞
		v.	伞降；空投
strikingly	/'straɪkɪŋli/	adv.	显著地，突出地，引人注目地
documentation	/ˌdɒkjumen'teɪʃn/	n.	证明文件；归档；必备资料；文件记载
by the name of			名叫……的

137

eco-friendly	/ˌiːkəʊˈfrendli/	adj.	对环境无害的；环保的
spoilt	/spɔɪlt/	adj.	变坏的，变质的，腐败的
insulate	/ˈɪnsjuleɪt/	v.	使隔热
sawdust	/ˈsɔːdʌst/	n.	锯木屑
circulate	/ˈsɜːkjəleɪt/	v.	循环
be lined with			排列着

138

barber	/ˈbɑːbə(r)/	n.	（为男子理发、修面的）理发师
spin	/spɪn/	v.	纺纱；纺线
workable	/ˈwɜːkəbl/	adj.	可行的；行得通的；
strand	/strænd/	n.	（线、绳、金属线，毛发等的）股，缕
water frame			水力纺纱机
team up with			和……协作

139

razor	/ˈreɪzə(r)/	n.	剃须刀
clam	/klæm/	n.	蛤；蛤蜊；蚌
conceptualise	/kənˈseptʃuəlaɪz/	v.	构思；使形成观念；将……概念化（为……）
double-edged	/ˌdʌb(ə)l edʒd/	adj.	双刃的
disposable	/dɪˈspəʊzəb(ə)l/	adj.	用后即丢弃的；一次性的
burial chambers			墓室
mass produce			大量生产

140

mayonnaise	/ˌmeɪəˈneɪz/	n.	蛋黄酱（用作三明治、色拉等的调味品）
culinary	/ˈkʌlɪnəri/	adj.	烹饪的；食物的
frequently	/ˈfriːkwəntli/	adv.	频繁地；时常；不断地
emulsion	/ɪˈmʌlʃn/	n.	乳剂；乳状液
conquer	/ˈkɒŋkə(r)/	v.	征服；克服；战胜；攻克
lack	/læk/	v.	缺乏；短缺
deli	/ˈdeli/	n.	熟食店
blue ribbon	/ˈbluːˈrɪbən/	n./adj.	最高荣誉；一流的

141

accelerometer	/əkˌseləˈrɒmɪtə(r)/	n.	加速度计
velocity	/vəˈlɒsəti/	n.	高速；（沿某一方向的）速度
validate	/ˈvælɪdeɪt/	v.	确认；证实；确证
Newtonian	/njuːˈtəʊniən/	adj.	牛顿学说的
mass	/mæs/	n.	质量
radius	/ˈreɪdiəs/	n.	半径范围；半径（长度）

Words and Expressions
（生词与短语）

142

carbonated	/ˈkɑːbəneɪtɪd/	adj.	含二氧化碳的
sparkling	/ˈspɑːklɪŋ/	adj.	起泡的
seltzer	/ˈseltzə(r)/	n.	塞尔脱兹（含汽）矿泉水
fizzy	/ˈfɪzi/	adj.	起泡的
mineral	/ˈmɪnərəl/	adj.	矿物（性）的；含矿物的
pressurised	/ˈpreʃəˌraɪzd/	adj.	增压的；加压的
flavoured	/ˈfleɪvəd/	adj.	有……味道的；添加了味道的

143

textile mill	/ˈtekstaɪl mɪl/	n.	纺织厂
originally	/əˈrɪdʒənəli/	adv.	原来；起初
corresponding	/ˌkɒrɪˈspɒndɪŋ/	adj.	符合的；相应的；相关的
weave	/wiːv/	n.	编法；织法
roving	/ˈrəʊvɪŋ/	n.	粗纱；粗纺
decrease	/dɪˈkriːs/	v.	降低；减少
labour	/ˈleɪbə(r)/	n.	劳动力
warp	/wɔːp/	n.	（织布机上的）经纱
weft	/weft/	n.	（织布机上的）纬纱
spinning jenny			珍妮纺纱机；（初期的）多轴纺纱机

144

matzo	/ˈmætsəʊ/	n.	无酵饼（犹太人在逾越节时吃）
conception	/kənˈsepʃn/	n.	构想；构思
earl	/ɜːl/	n.	伯爵
avid	/ˈævɪd/	adj.	热衷的；酷爱的；渴望的；渴求的
gambler	/ˈɡæmblə/	n.	赌徒
streak	/striːk/	n.	（尤指体育比赛或赌博中）的运气，手气
ongoing	/ˈɒnɡəʊɪŋ/	adj.	持续存在的；仍在进行的；不断发展的
utensil	/juːˈtensl/	n.	器皿；家什
bitter herb			苦草
attribute to			把……归因于

145

smallpox	/ˈsmɔːlpɒks/	n.	天花
vaccine	/ˈvæksiːn/	n.	疫苗
inoculate	/ɪˈnɒkjuleɪt/	v.	（给……）接种，打预防针
cowpox	/ˈkaʊˌpɒks/	n.	牛痘
virus	/ˈvaɪrəs/	n.	病毒
variola	/vəˈraɪələ/	n.	（医）天花
affect	/əˈfekt/	v.	使感染
mammal	/ˈmæml/	n.	哺乳动物

cattle	/ˈkætl/	n.	牛
contract	/kɒnˈtrækt/	v.	感染
deliberately	/dɪˈlɪbərətli/	adv.	有意地；蓄意地
side effect	/saɪd ɪˈfekt/	n.	副作用
vaccination	/ˌvæksɪˈneɪʃn/	n.	接种疫苗
variolation	/ˌveərɪəˈleɪʃn/	n.	人痘接种；天花接种；引痘

146

paddle	/ˈpædl/	n.	桨；船桨
flop	/flɒp/	v.	移动
colleague	/ˈkɒliːɡ/	n.	同事；同行；同僚
demonstration	/ˌdemənˈstreɪʃn/	n.	示范；演示

147

cast	/kɑːst/	v.	浇铸；铸造
erect	/ɪˈrekt/	v.	建立；建造
rib	/rɪb/	n.	（船或屋顶等的）肋拱
discrepancy	/dɪsˈkrepənsi/	n.	差异；不符合；不一致
identical	/aɪˈdentɪkl/	adj.	完全同样的；相同的；同一的

148

magnificent	/mæɡˈnɪfɪsnt/	adj.	值得赞扬的
aloft	/əˈlɒft/	adv.	在高空中
tether	/ˈteðə(r)/	n.	（拴牲畜的）拴绳
ascent	/əˈsent/	n.	上升；登高；升高
impressive	/ɪmˈpresɪv/	adj.	令人印象深刻的；可观的

149

thresher	/ˈθreʃə/	n.	脱粒机；打谷机；脱谷机；打叶机
stalk	/stɔːk/	n.	秆；柄
husk	/hʌsk/	n.	外皮
flail	/fleɪl/	n.	梿枷（旧时长柄脱粒农具）
time-consuming	/taɪm kənˈsjuːmɪŋ/	adj.	费时的；耗时间的
hand-fed	/ˈhænd fed/	adj.	人工进料的
upright	/ˈʌpraɪt/	adj.	直立的；挺直的；竖直的；垂直的
portable	/ˈpɔːtəbl/	n.	便携机
traction engine			牵引机车
separate ... from ...			把……分离

150

cotton gin	/ˈkɒtn dʒɪn/	n.	轧花机；轧棉机
boll	/bəʊl/	n.	棉铃
wire tooth			金属齿针
comb out			梳出；清理

Words and Expressions
（生词与短语）

151

ball bearing	/bɔːl ˈbeərɪŋ/	n.	滚珠轴承
grooved	/gruːvd/	adj.	有沟的；有槽的
friction	/ˈfrɪkʃn/	n.	摩擦；摩擦力
possess	/pəˈzes/	v.	拥有；控制；具有（特质）
capacity	/kəˈpæsəti/	n.	容量；生产能力；容积；功率
alignment	/əˈlaɪnmənt/	n.	排成直线
crucial	/ˈkruːʃ(ə)l/	adj.	至关重要的；关键性的
resilient	/rɪˈzɪliənt/	adj.	可迅速恢复的；有适应力的；有弹性（或弹力）的；能复原的
fabric-reinforced	/ˈfæbrɪk ˌriːɪnˈfɔːst/	adj.	以纤维加强的
phenolic	/fɪˈnɒlɪk/	adj.	（化）酚的
resin	/ˈrezɪn/	n.	树脂；合成树脂

152

reaction	/riˈækʃn/	n.	反应；化学反应
feather	/ˈfeðə(r)/	n.	羽毛；翎毛
spark	/spɑːk/	n.	火花；火星
pistol	/ˈpɪstl/	n.	手枪
jar	/dʒɑː(r)/	n.	（玻璃）罐子；广口瓶
methane	/ˈmiːθeɪn/	n.	甲烷；沼气
revolve	/rɪˈvɒlv/	v.	旋转；环绕；转动
constant flow			定量流动

153

protractor	/prəˈtræktə(r)/	n.	量角器；分度规
angle	/ˈæŋgl/	n.	角；角度
astronomical	/ˌæstrəˈnɒmɪkl/	n.	天文学的；天文的
semicircular	/ˌsemiˈsɜːkjələ(r)/	adj.	半圆形的
multifaceted	/ˌmʌltiˈfæsɪtɪd/	adj.	多方面的；要从多方面考虑的
navigational	/ˌnævɪˈgeɪʃənəl/	adj.	航行的；飞行的
naval	/ˈneɪvl/	adj.	海军的
rotatable	/rəʊˈteɪtəbl/	adj.	可旋转的
marine	/məˈriːn/	adj.	海船的；海上贸易的
angular	/ˈæŋgjələ(r)/	adj.	有棱角的；有尖角的
navigator	/ˈnævɪgeɪtə(r)/	adj.	（飞机、船舶等上的）航行者，航海者
be relative to			与……有关；随……转移

154

hang glider	/hæŋ ˈglaɪdə(r)/	n.	悬挂式滑翔机
suspend	/səˈspend/	v.	挂
hand gliding			滑翔翼运动

drift down			在巡航中下降

155

quinine	/kwɪ'ni:n/	n.	奎宁
cinchona	/sɪŋ'kəʊnə/	n.	金鸡纳树
malaria	/mə'leəriə/	n.	疟疾
shiver	/'ʃɪvə/	v.	颤抖

156

merchant	/'mɜːtʃnt/	n.	商人；批发商；（尤指）进出口批发商
commercial	/kə'mɜːʃl/	adj.	赢利的；以获利为目的的
previous	/'priːviəs/	adj.	先前的；以往的；（时间上）稍前的
machine-stamped	/mə'ʃiːn stæmpt/	adj.	凸版印刷机式的

157

solar cell	/ˌsəʊlə(r) 'sel/	n.	太阳能电池
convert	/kən'vɜːt/	v.	转换；可转变为；可变换成
photovoltaics	/ˌfəʊtəʊvɒl'teɪɪks/	n.	光电学
electrode	/ɪ'lektrəʊd/	n.	电极
electrolyte	/ɪ'lektrəlaɪt/	n.	电解质；电解液
solution	/sə'luːʃn/	n	溶液
voltage	/'vəʊltɪdʒ/	n.	电压；伏特数
panel	/'pænl/	n.	镶板
conversion	/kən'vɜːʃn/	n.	转换；转化；转变

158

tractor	/'træktə/	n.	拖拉机；牵引机；（牵引式挂车的）牵引车
coin	/kɔɪn/	v.	创造；提出
gasoline	/'gæsəliːn/	n.	（美）汽油
thresh	/θreʃ/	v.	（用机器）使脱粒
back and forth			来回地
backwards and forwards			来回地

159

reaper	/'riːpə/	n.	收割机
horse-drawn	/hɔːs drɔːn/	adj.	马拉的
manual	/'mænjʊəl/	adj.	用手的；手工的；手动的
scythe	/saɪð/	n.	长柄大镰刀
sickle	/'sɪk(ə)l/	n.	镰刀
acre	/'eɪkə/	n.	英亩（4,840平方码，约为4,050平方米）
rake	/reɪk/	n.	耙子；耙状工具
export	/ɪk'spɔːt/	v.	出口
progress	/prəʊ'gres/	v.	进展；进步；前进；改进
establishment	/ɪ'stæblɪʃmənt/	n.	机构

Words and Expressions
（生词与短语）

160

stethoscope	/'steθəskəʊp/	n.	听诊器
procedure	/prə'siːdʒə(r)/	n.	步骤；手术
plump	/plʌmp/	adj.	丰腴的；微胖的；丰满的
whereby	/weə(r)'baɪ/	adv.	借以；凭此；由于

161

philosopher	/fɪ'lɒsəfə/	n.	哲学家；深思的人；善于思考的人
pinhole	/'pɪnˌhəʊl/	n.	针孔
inverted	/ɪn'vɜːtɪd/	adj.	倒置的；反转的
permanent	/'pɜːmənənt/	adj.	永久的，永恒的，长久的
capture	/'kæptʃə/	v.	拍摄
bitumen	/'bɪtʃəmən/	n.	沥青
layman	/'leɪmən/	n.	外行；门外汉；非专业人员

162

scooter	/'skuːtə(r)/	n.	（儿童）滑板车
carriage	/'kærɪdʒ/	n.	四轮马车
tyre	/'taɪə(r)/	n.	轮胎
sprocket	/'sprɒkɪt/	n.	链轮
steering bar			导向杆
gear system			齿轮传动装置

163

suspender	/sə'spendə(r)/	n.	吊裤带；背带［常用复数suspenders］
ribbon	/'rɪbən/	n.	丝带；（用于捆绑或装饰的）带子
clasp	/klɑːsp/	n./v.	（包、皮带或首饰的）搭扣；扣住
waistband	/'weɪstbænd/	n.	裤腰

164

strike	/straɪk/	v.	擦，划（火柴）；击出（火星）
flame	/fleɪm/	n.	火焰
pinewood	/'paɪnwʊd/	n.	松树；松木
combustible	/kəm'bʌstəbl/	adj.	易燃的；可燃的
starch	/stɑːtʃ/	n.	淀粉；（浆衣服、床单等用的）淀粉浆
Swedish	/'swiːdɪʃ/	adj.	瑞典（式）的；瑞典人（语）的

165

Scottish	/'skɒtɪʃ/	adj.	苏格兰的；苏格兰人的
cement	/sə'ment/	v.	（用水泥、胶等）黏结，胶合
puncture	/'pʌŋktʃə(r)/	v.	在……上扎孔（或穿孔）；（被）刺破
seam	/siːm/	v.	缝合；接合；缝拢
deteriorate	/dɪ'tɪəriəreɪt/	v.	变坏；恶化；退化
vulcanise	/'vʌlkənaɪz/	v.	（橡胶等）硫化；硬化

166

savant	/ˈsævnt/	n.	博学之士；学者；专家
secure	/sɪˈkjʊə(r)/	v.	（尤指经过努力）获得，取得，实现
establish	/ɪˈstæblɪʃ/	v.	建立；创立；设立
omnibus	/ˈɒmnɪbəs/	n.	公共汽车
toll gate	/ˈtəʊl geɪt/	n.	道路收费卡口；收费站；收费口；关卡
turnpike	/ˈtɜːnpaɪk/	n.	收费公路
pick up			（汽车，飞机）接客
set down			下客

167

braille	/breɪl/	n.	布拉耶盲文（凸点符号）
sufficient	/səˈfɪʃnt/	adj.	足够的；充足的
infection	/ɪnˈfekʃn/	n.	传染病；感染
trace	/treɪs/	v.	追踪；勾画出（轮廓）
struggle	/ˈstrʌɡl/	v.	奋斗；努力；争取
master	/ˈmɑːstə(r)/	v.	精通；掌握
emboss	/ɪmˈbɒs/	v.	压印浮凸字体（或图案）；凹凸印
alphanumeric	/ˌælfənjuˈmerɪk/	adj.	含有字母和数字的；字母与数字并用的
punch	/pʌntʃ/	v.	给……打孔；按（键）；压（按钮）
slate	/sleɪt/	n.	板岩；石板
visually	/ˈvɪʒuəli/	adv.	视觉上地
be in place			已经到位；形成

168

sewing	/ˈsəʊɪŋ/	n.	缝纫
canvas	/ˈkænvəs/	n.	帆布
chain stitch	/tʃeɪn stɪtʃ/	n.	链状绣；链式针法
disparate	/ˈdɪsprət/	adj.	由不同的人（或事物）组成的；不相干的
innovation	/ˌɪnəʊˈveɪʃn/	n.	创造；创新；改革
botch	/bɒtʃ/	v.	笨拙地弄糟（某事物）
due	/djuː/	adj.	应有的
recognition	/ˌrekəɡˈnɪʃn/	n.	承认；认可

169

pouch	/paʊtʃ/	n.	小袋子；荷包
onset	/ˈɒnˌset/	n.	开端，发生
harness	/ˈhɑːnɪs/	n.	马具
saddle	/ˈsædl/	n.	马鞍
packer	/ˈpækə(r)/	n.	包装工
Parisian	/pəˈrɪziən/	adj.	巴黎的，巴黎式的；巴黎人的
with the onset of			随着……的来临

Words and Expressions
（生词与短语）

170

combine harvester	/kəm'baɪn 'hɑːvɪstə(r)/	n.	联合收割机
header	/'hedə(r)/	n.	割穗机
predecessor	/'priːdɪsesə(r)/	n.	原先的东西；被替代的事物
impact	/'ɪmpækt/	n.	巨大影响；强大作用
advanced	/əd'vɑːnst/	adj.	先进的
profitable	/'prɒfɪtəbl/	adj.	有利润的；赢利的
individually	/ˌɪndɪ'vɪdʒuəli/	adv.	分别地；单独地；各别地
have a ... impact on ...			对……产生了……影响（作用）

171

telegraph	/'telɪˌɡrɑːf/	n.	电报（通信方式）
transmit	/trænz'mɪt/	v.	传送；输送
dit	/dɪt/	n.	电码的点号
dah	/dɑː/	n.	无线电或电报电码中的一长划
dash	/dæʃ/	n.	破折号
interval	/'ɪntəvl/	n.	（时间上的）间隔，间隙，间歇
standard	/'stændəd/	adj.	普通的；正常的；通常的；标准的
depend on			依靠；取决于
lay the groundwork for			为……打下基础

172

semaphore	/'seməˌfɔː(r)/	n.	信号标；旗语
optical	/'ɒptɪkl/	adj.	光学的
electrolysis	/ɪˌlek'trɒləsɪs/	n.	电解作用
paper tape	/peɪpə(r) teɪp/	n.	（老式计算机记录数据的）打孔纸带
electromechanical	/ɪ'lektrəʊmɪ'kænɪkəl/	adj.	电机的；电机学的
interpret	/ɪn'tɜːprɪt/	v.	诠释；说明
corporation	/ˌkɔːpə'reɪʃn/	n.	（大）公司

173

postage	/'pəʊstɪdʒ/	n.	邮资；邮费
portrait	/'pɔːtreɪt/	n.	肖像；描绘；半身画像；半身照
profile	/'prəʊfaɪl/	n.	轮廓；形象；外形

174

suspension bridge	/sə'spenʃn brɪdʒ/	n.	吊桥，悬索桥
visualise	/'vɪʒuəlaɪz/	v.	想象；构思；设想
consent	/kən'sent/	v.	同意；允许；准许
jointly	/dʒɔɪntli/	adv.	连带地；共同地；联合地
commission	/kə'mɪʃn/	v.	正式委托（谱写、制作、创作或完成）
span	/spæn/	n.	持续时间；范围；跨度
navigable	/'nævɪɡəbl/	adj.	可航行的；适于通航的

otherwise	/ˈʌðəwaɪz/	adv.	否则；不然
pier	/pɪə(r)/	n.	桥墩
drove	/drəʊv/	n.	畜群

175

stapler	/ˈsteɪplə(r)/	n.	订书机
fastener	/ˈfɑːsnə(r)/	n.	扣件
clinch	/klɪntʃ/	v.	钉牢；钉住
staple	/ˈsteɪpl/	n.	订书钉

176

astringent	/əˈstrɪndʒnt/	n.	止血剂；收敛剂
dress	/dres/	v.	清洗；包扎；敷裹（伤口）
dye	/daɪ/	n.	染料；染液
alum	/ˈæləm/	n.	明矾；硫酸铝；十二水合硫酸铝钾
acknowledge	/əkˈnɒlɪdʒ/	v.	承认；认识
Danish	/ˈdeɪnɪʃ/	adj.	丹麦的
commercially	/kəˈmɜːʃəli/	adv.	商业上；商业地；商业上地
metal base			金属基材

177

voltmeter	/ˈvəʊltmiːtə(r)/	n.	电压表；伏特计
circuit	/ˈsɜːkɪt/	n.	电路；线路
galvanometer	/ˌgælvəˈnɒmɪtə/	n.	电表；电流计
conductor	/kənˈdʌktə(r)/	n.	导体
needle	/ˈniːdl/	n.	针
coil	/kɔɪl/	n.	绕组；线圈
consistent	/kənˈsɪstənt/	adj.	一致的；始终如一的；连续的
surname	/ˈsɜːneɪm/	n.	姓
jerk	/dʒɜːk/	v.	猛拉
electrical potential			电位；电势
have one's root in			根植于

178

dirigible	/ˈdɪrɪdʒəbl/	n.	飞艇；汽艇
cigar-shaped	/sɪˈgɑː(r) ʃeɪpt/	adj.	雪茄烟形的
propulsion	/prəˈpʌlʃn/	n.	推进；推动力
gondola	/ˈgɒndələ/	n.	（热气球、飞船上的）吊舱，吊篮
accommodate	/əˈkɒmədeɪt/	v.	容纳；提供住宿（或膳宿、座位等）
cargo	/ˈkɑːgəʊ/	n.	（船或飞机装载的）货物
steering mechanism			转向机械；操舵机构

179

| platinum | /ˈplætɪnəm/ | n. | 铂；白金 |

Words and Expressions
（生词与短语）

| mould | /məʊld/ | v. | 铸；形成 |
| poke | /pəʊk/ | v. | 戳；捅 |

180

synchronise	/'sɪŋkrənaɪz/	v.	同速进行
pendulum	/'pendjələm/	n.	钟摆
synchronisation	/ˌsɪŋkrənaɪ'zeɪʃn/	n.	同步性；同步画面
revision	/rɪ'vɪʒn/	n.	修订；修改
pantelegraph	/pæn'teləgrɑːf/	n.	（早期的）有线传真电报

181

| hoist | /hɔɪst/ | n. | 起重机；吊车；（残疾人用）升降机 |
| counterbalance-type | /ˌkaʊntə'bæləns taɪp/ | adj. | 平衡重式的 |

182

anaesthesia	/ˌænəs'θiːziə/	n.	麻醉
administer	/əd'mɪnɪstə(r)/	v.	施行；执行
unconsciousness	/ʌn'kɒnʃəsnəs/	n.	昏迷；无知觉状态
condition	/kən'dɪʃn/	n.	健康状况
render	/'rendə(r)/	v.	使成为；使变得；使处于某状态
sensation	/sen'seɪʃ(ə)n/	n.	感觉；知觉
come about			发生，形成，到来

183

Hungarian	/hʌŋ'ɡeəriən/	adj.	匈牙利的
obstetrician	/ˌɒbstə'trɪʃn/	n.	产科医生
ward	/wɔːd/	n.	病房；病室
midwife	/'mɪdwaɪf/	n.	助产士；接生员
dissection	/dɪ'sekʃn/	n.	解剖
corpse	/kɔːps/	n.	尸体；死尸；尸首
sanitise	/'sænɪtaɪz/	v.	（用化学制剂）消毒，使清洁
diluted	/daɪ'luːtɪd/	adj.	稀释的；淡的；稀薄的
surgeon	/'sɜːdʒn/	n.	外科医生

| bleaching powder | | | 漂白粉 |

Index
（索引）*

A

a length of   23
a stretch of   75
abrasiveness   43
abstract   9
abundance   21
accelerometer   141
accessory   31
accommodate   178
accompany   112
accredited   120
accumulate   128
acknowledge   176
acquire   5
acre   159
adhesive   5
administer   182
adorn   38
advanced   170
affect   145
aggregate   50
aid   127
air pressure   134
air tube   130
alchemist   104
alcohol   19
alcohol thermometer   126
alcoholic   19
align   30
alignment   151
alloy   35
almond   60
aloft   148
alphanumeric   167

alum   176
amber   113
ammunition   122
anaesthesia   182
ancestor   1
angle   153
angular   153
anklets   93
annual   41
answer nature's call   49
ape-like   1
apex   9
apply for   132
approximate   125
approximately   50
approximation   125
aqueduct   50
Arab   104
arch   61
archaeological   67
archaeologist   2
archaeology   33
archer   8
Archimedes' screw   68
Arctic   6
aristocracy   120
aromatic   60
arrow   8
artery   135
artillery   67
ascent   148
ash   34
aspect   97
associate   19
Assyrian   56

astringent   176
astrologer   99
astronomer   99
astronomical   153
astronomy   132
attach   23
attach to   108
attempt   18
attire   92
attribute to   144
augment   99
aurochs   9
auspicious   84
authentic   118
avid   144
awkward   49
axe   23
axle   30

B

Babylonian   99
back and forth   158
back then   4
backpack   84
backwards and forwards   158
bacteria   15
ball bearing   151
ballista   76
ballpoint pen   39
barber   138
bare   93
bare socks   93
bark   25
barometer   134
barrel   77

*本索引中单词或词组后的数字为其首次出现的篇目号，供查阅用。

Index
（索引）

barter　22
be adopted by　120
be composed of　83
be considered to be　25
be derived from　134
be devoted to doing something　115
be equipped with　84
be frowned upon　59
be given the credit for ...　98
be in place　167
be lined with　137
be reduced to　107
be relative to　153
betel nut　43
be traced back to　89
beam　86
beeswax　23
beneficial　127
bergamot　60
beverage　19
binder　48
binding substance　34
bireme　32
bisect　99
bison　9
bitter herb　144
bitumen　161
blacksmith　45
blade　3
blazing　51
bleaching powder　183
blend　93
bloomers　92
bloomery　42
blue ribbon　140
blurry　119
bobbin　108
bodice　117
boll　150
bolt　112

bone china　118
bore　101
botch　168
bow　8
bow drill　46
bow tie　129
boxwood　10
braid　13
braided　31
braille　167
brass　10
break down　19
brew　118
bribe　71
brick　7
bridal　84
bristle　96
brittle　114
bucket　25
buckram　117
bullet　109
bumpy　47
burial chambers　139
burst　109
bury　15
by the name of　136

C

calamus　60
calliper　74
canal　24
candlestick　85
candlewick　108
cannon　101
canvas　168
capacity　151
capture　161
carbonated　142
card　108
cargo　178

carpentry　23
carriage　162
carry forward　87
cart　127
cartridge　122
carve　20
carving　16
cast[1]　29
cast[2]　147
catapult　76
catch on　120
cater to　70
cattle　145
cautious　5
cave　8
cavity　43
ceiling　56
cellulose pulp　94
cement[1]　34
cement[2]　165
cement mix　34
cemetery　47
ceramic[1]　2
ceramic[2]　39
chain stitch　168
chalk　43
chamber　90
channel[1]　24
channel[2]　41
character　116
charcoal　43
chariot　64
charioteers　64
charred　2
chest　118
churn　81
cigar-shaped　178
cinchona　155
cinnamon　85
circa　51

circuit 177
circular 99
circulate 137
circulation 135
civil 89
civilisation 7
claim 132
clam 139
clasp 163
clay soil 17
clay tablet 18
clinch 175
clinical 126
clotting factor 135
coarse 50
cobalt 14
codex 27
coil 177
coin 158
collapsible 43
collar 129
colleague 146
column 56
comb out 150
comb with 108
combine harvester 170
combustible 164
come about 182
come into existence 53
come into fashion 121
come into the picture 77
come up with 8
command-vehicles 64
commercial 156
commercially 176
commission 174
commoner 70
compact 50
compass^1 74
compass^2 83

complement^1 102
complement^2 129
complex 5
component 31
compose 83
compound 80
comprise 127
conceal 82
concept 24
conception 144
conceptualise 139
concrete 50
condensed 50
condenser 131
condition 182
conductor 177
cone 101
conquer 140
consent 174
considerable 80
consistent 177
constant 37
constant flow 152
constellation 83
constitute 124
consume 57
contain 122
contract^1 126
contract^2 145
contraption 30
contrary 132
contrary to 132
contribute 25
contribution 14
controversial 6
conversion 157
convert 157
convex^1 111
convex^2 119
convey 91

coolant 40
copper 23
core^1 1
core^2 122
cork 112
cork oak 128
corporation 172
corpse 183
correspond 47
corresponding 143
corset 117
cosmetic 40
cottage 128
cotton gin 150
coulter 45
counterbalance-type 181
counterpart 32
courier 91
course 89
covet 118
cowpox 145
crane-like 41
crater 132
cravate 133
crawl 20
credit^1 98
credit^2 107
Croatian 133
crossbow 8
crucial 151
crucial 2
crude 3
crush 14
crystal 111
cubit 62
cuisine 58
culinary 140
cultivation 128
cumbersome 100
cuneiform script 18

Index（索引）

curcus publicus   91
curiosity   106
currency   22
current   81
currently   93
curved   36
cylinder   98
cylindrical   98

D

dah   171
dam   24
Danish   176
dash   171
date back to   2
dawn¹   23
dawn²   99
debatable   133
debt   36
decade   21
decay   46
decipher   18
deck   84
declare   125
decor   110
decrease   143
define   19
dehydrate   15
deli   140
deliberately   145
delicate   110
demonstration   146
denote   100
dense   105
density   98
dental   46
dentistry   46
depend on   171
depict   99
deposit¹   81

deposit²   100
deposit³   114
derive   29
derive from   29
despite   37
deteriorate   165
determine   89
device   73
devise   24
diameter   82
differentiate   1
diluted   183
dimensional   99
diminish   56
dip   40
dip into   40
dirigible   178
discharge   103
discrepancy   147
discus   30
disparate   168
disposable   139
dispute   2
dissection   183
dissolve   37
distill   60
distinct   38
distracted   102
distribute   56
dit   171
divert   24
document   27
documentation   136
dome   87
double-edged   139
downward   75
drain   49
drain out   68
drainage   61
drastically   6

draw   127
drawback   62
drawers   92
dress   176
drift down   154
drove   174
dry out   15
dry up   17
dual   38
due   168
due to   106
durability   112
durable   25
Dutch   118
dye   176

E

earl   144
eco-friendly   137
egg whites   20
elaborate   38
elaborately   20
elbow   62
electrical potential   177
electrode   157
electrolysis   172
electrolyte   157
electromechanical   172
element   32
eliminate   39
elite   70
embed   83
emblem   102
emboss   167
emerge   106
empty into   82
emulsion   140
enclose   84
enclosed   126
encounter   91

engineer 5
enhance 14
ensure 79
enthusiast 118
entire 80
era 1
erect 147
error 125
establish 166
establishment 159
estimate 2
eulachon 85
Eurasian 8
evaporate 50
evenly 116
eventually 3
evidence 2
evil 40
evolution 2
evolve 1
evolve into 121
excavate 47
exclusively 118
expand 126
explode 104
explosive 104
explosive charge 101
export 159
expose 25
exposure 92
exquisite 118
extensive 99
extensively 61
extract[1] 22
extract[2] 46
extract[3] 60
extremely 105
eyepiece 119

F

fabric 36
fabric-reinforced 151
fade 35
fade away 35
fair skin 63
fairly 105
fall into 37
fall out of 75
fancy 110
fashion statement 31
fasten 112
fastener 175
faucet 82
feat 5
feather 152
feel like doing 16
felt pen 39
ferment 19
fermentation 19
fervour 60
fibre[1] 16
fibre[2] 37
fight over 22
filter 60
fine 88
finely 50
fire lance 103
firearm 109
fit into 36
fizzy 142
flail 149
flake 1
flame 164
flange 127
flanged 127
flatten 99
flavour 102
flavoured 142

flaw 17
fleecy 108
flesh 15
flexible 25
fling 76
flint 109
flintlock 109
float 130
flop 146
flourish 49
fluffy 20
fluoride 43
flush 49
folk 57
for instance 8
force of gravity 86
forceps 46
formerly 109
formidable 8
formula[1] 25
formula[2] 35
fortune teller 83
fountain pen 39
fragile 77
frayed 96
frequently 140
friction 151
fro 77
frown 59
fulcrum 86
full-fledged 25
function 111
furnace 42
furthermore 17

G

gain a lot of popularity 110
galvanometer 177
gambler 144
garment 117

Index（索引）

H/I

gasket 130
gasoline 158
gauge 86
gauntlet 71
gear system 162
generate 2
get soaked into 17
gild 84
ginseng 43
give the credit to 110
glass globe 78
glaze 102
goffering iron 124
gondola 178
gown 107
grab 16
gradually 107
grain 25
graphite 114
grass leather 13
grave 47
gravel 48
gravity 75
grind 88
grip 127
groove 81
grooved 151
gum 43
gunsmith 109

H

hammer 42
hand 115
hand gliding 154
hand tools 13
hand-chiselled 5
hand-fed 149
hang glider 154
Harappan 54
harness 169
harvest 128
haul 80
have a ... impact on ... 170
have one's root in 177
header 170
headgear 37
heat-resistant 118
helical groove 109
herbal mint 43
herd 47
hide 23
hint at 16
hire 124
hoe 59
hoist[1] 80
hoist[2] 181
hold up 87
hollow 39
home decor 110
hook[1] 11
hook[2] 111
horizontal 29
horn 8
horoscope 99
horse-drawn 159
hose 121
hosiery 121
hull 32
Hungarian 183
hurdle 34
husk 149
hydraulic 34
hydrometer 98

I

iconic 31
identical 147
identify 69
igloo 87
ignite 109
immense 22
immigrant 37
immortality 104
impact 170
imperial court 94
implement 42
importer 118
impressive 148
improvisation 61
improvise 70
in a ... manner 126
in a true sense 48
in comparison to 77
in prayer 102
in sync with 89
in use 83
incriminate 71
indicate 25
indicator 44
indigo 55
individual 100
individually 170
infantry 64
infection 167
initial 25
initially 62
innovate 32
innovation 168
innovative 57
inoculate 145
inscribe 27
inscription 74
insert ... into 122
inspire 20
install 54
instrument 10
insulate 137
insulated 90
integral 45
intellectual 1

intermediate   67
internal   101
interpret   172
interval   171
intricate   70
inverted   161
involve   19
irrigation   24
ivory   11

J/K/L/M

J

jagged   33
jar   152
jaw   33
jerk   177
jet   82
jewelled   31
jointly   174
juniper   114

K

keep ... at bay   15
keep a count of   18
keep account of   115
keep off   20
kiln   7
kiln-fired brick   7
kinetic   81
knit   93
knot   133
kohl   40

L

laboratory   51
labour   143
labourer   135
lack   140
land fill   88
landscape   84
Latin   60
latitude   99

latrine   49
launch¹   106
launch²   130
lavatory   49
lavish   57
lay   111
lay the foundation   13
lay the groundwork for   171
layman   161
lead   70
league   4
leap month   89
leave aside   17
leftover   17
legacy   87
legend   37
legend has it that   37
leggings   93
lens   119
lever   41
liberally   40
limekiln   50
line   17
linear accerleration   141
linen   71
liquid-propellant   106
literacy   116
literary   27
literature   16
litter   84
load   76
loaded   131
locker room   70
locksmith   115
locomotive   127
lodestone   83
log   4
longbow   8
longevity   14
longitude   89

loom   105
loop   36
loose   27
loosely   108
loosen   112
lower class   38
lunisolar   89
luxury   49
lye   102

M

machine-stamped   156
magnetised   83
magnetism   113
magnetometer   83
magnification   111
magnificent   148
magnifier   111
magnifying glass   78
maintenance   75
make a profit   111
make an appearance   110
makeshift   51
malaria   155
mammal   145
mammoth   6
mandarin   84
manual   159
manually   81
manufacture   105
marble   72
marine   153
masculine   120
masonry   7
mass   141
mass produce   139
master   167
matted   93
matzo   144
maximum   119

Index（索引）

N/O/P

mayonnaise 140
mechanisation 116
mechanism 41
medical practice 135
medieval 25
Mediterranean 8
merchant 156
mercury 98
Mesopotamian 7
metal base 176
metallic 100
metallurgy 23
methane 152
microorganism 15
midwife 183
migrate 4
military 8
mill 81
millennium 18
milliner 123
milling 88
mineral[1] 83
mineral[2] 142
minimal 14
minimum 119
Minoan 54
minute 74
misidentify 114
modification 30
moist 94
molecule 37
monk 102
monument 61
moon 132
mortar[1] 21
mortar[2] 88
motion 81
motivate 102
mould[1] 7
mould[2] 179

mouldboard 45
mounted 8
movable 116
mulberry bark 94
mulbery 37
multifaceted 153
musk ox 6
muslin 40
myrtle 60
mythology 33

N

name ... after ... 113
narrow tube 132
natural selection 1
naval 153
navigable 174
navigation 99
navigational 153
navigator 153
navy 130
necessity 51
needle 177
negative lens 132
Newtonian 141
nib 39
nilometer 24
nobility 63
nomad 47
normally 103
novelty 119
nuclear-powered 130
numerical scale 126
nut 112
nylon 93

O

oar 32
objective 119
observation 28
obsess 99

obsolete 18
obstacle 35
obstetrician 183
obtain 3
occasional 71
occur 9
Oldowan 1
olive 112
omnibus 166
on a lange scale 23
ongoing 144
onset 169
openwork 110
optical 172
optician 111
option 9
ore 42
orient 83
origin 91
originally 143
originate 9
ornament[1] 71
ornament[2] 84
ornate 56
otherwise 174
overflow 24
overlay 107
oyster shell 43

P

packer 169
paddle 146
palanquin 84
palette 9
palm 38
panel 157
pantelegraph 180
paper tape 172
papyrus 27
parachute 136

parasol   63
Parisian   169
particle   66
particular   126
pashmina   69
paste[1]   21
paste[2]   57
pastoralism   47
pastry   57
patent   132
path-breaking   22
pattern   9
pave   35
pave the path for   35
paved road   48
peak   64
peasant   107
pebble   50
pedestal   72
penalty   37
pendulum   180
perish   15
perishable   28
permanent   161
Persian   120
pestle   88
petite   120
pharaoh   48
phase[1]   9
phase[2]   89
phenolic   151
philosopher   161
photovoltaics   157
physician   113
pick up   166
pier   174
pigment   14
pigmentation   14
pillar   56
pillow   20

pin   73
pinewood   164
pinhole   161
pioneer[1]   87
pioneer[2]   106
pipe   49
pistol   152
piston   131
pit   90
pith   94
pivot   83
plain   69
plantation   59
plasma   135
platelet   135
platinum   179
playwright   78
pleat   124
plier   51
plot   99
plough   45
ploughshare   45
pluck   36
pluck off   36
plug   128
plumbing   72
plump   160
pointed   122
poke   179
Polish   18
polish   119
political affiliation   123
porcelain   46
portable[1]   115
portable[2]   149
portrait   173
positive lens   132
possess   151
possess   12
postage   173

porter   84
potter   30
pottery   17
pouch   169
powdered   43
power generation   81
pozzolana   50
practical   31
precisely   99
predecessor   170
pre-dynastic   90
prehistoric   9
prescription sunglasses   111
preservation   15
preservative   15
preserve   15
pressurised   142
prestige   57
pretzel   102
previous   156
previously   110
primarily   8
primitive   6
principal   116
principle   106
prior to   117
procedure   160
professional   106
profile   173
profitable   170
progress   159
projectile   101
projectile weapon   103
prolific   119
prominent   64
promote   116
promote literacy   116
prong   85
propel   109
propeller   130

Index (索引)

Q/R/S

property 135
propose 125
propulsion 178
prototype 103
protractor 153
pull out 37
pulley 80
pumice 43
pump 131
punch 167
puncture 165
purchase 130
purify 14

Q

quench-hardened 67
quicklime 50
quill 39
quinine 155

R

radius 141
rag 94
rake 159
range from 103
rank 63
razor 139
reaction 152
realisation 53
realise 37
realistic 99
reaper 159
reasoning 5
recipe 34
reclamation 88
recognition 168
recreation 92
recurved 8
redeem 100
reel 108
refill 41

refine 3
refined 22
refined sugar 66
refinement 116
regard … as … 116
region 4
relatively 85
relay 91
religious 27
remarkable 80
render 182
replace 127
replicate 23
repulse 58
resemblance 99
resemble 70
reserve 133
reservoir 82
reside 46
resident 114
resilient 151
resin 151
resist 52
respiration 19
rest 133
result in 5
retire 9
reveal 37
revision 180
revive 135
revolutionary 77
revolutionise 30
revolve 152
rib 147
ribbon 163
rice straw 94
ridiculous 77
right up to 107
rigid 111
rivet 111

robe 47
rock carving 16
rocketry 106
rod 50
roller ball pen 39
rot 25
rotary 131
rotate 115
rotor 81
round 109
roving 143
rowboat 130
royalty 63
rub 25
rudimentary 3
ruffled 92

S

sacrifice 53
saddle 169
sail 32
sandal 25
sandalwood 40
sanitation 70
sanitise 183
sap 26
savant 166
saw 33
sawdust 137
scale 98
scalp 38
scare 104
scooter 162
Scottish 165
scrape 9
scratch 39
screen 84
screw 112
screwdriver 112
scripture 27

U

unconsciousness 182
undergo 42
unearth 2
unglazed 118
unique 105
unravel 37
untimely 134
upper class 38
upright 149
utensil 144
utility 31
utmost 84

V

vaccination 145
vaccine 145
vacuum 134
validate 141
variola 145
variolation 145
vast 9
vault 97
vehicle 30
vein 135
velocity 141
velvet 105
vermillion 14
versatile 42
version 11
vertical 95
vertically 117
vessel 17
via 24
vinegar 15
violence 5
virus 145
visualise 174
visually 167
volatile 6
volcanic 34
voltage 157
voltmeter 177
voyage 99
vulcanise 165

W

wagonway 127
waist 92
waistband 163
ward 183
ward off 15
warp 143
warrior 69
washing soda 102
water closet 82
water frame 138
waterproof 32
water reed 13
wax 85
weave[1] 13
weave[2] 143
weft 143
weighing scale 44
whalebone 117
wheelbarrow 79
whereas 57
whereby 160
wick 40
wig 38
wildly 128
willow 16
wind[1] 23
wind[2] 115
wipe 69
wire tooth 150
wire-spoked wheel 30
with regards to 97
with respect to 125
with the onset of 169
withstand 118
witness 92
workable 138
woven 93
wrap 130
wrinkle 124
wrought 67

Y

yarn 108
yeast 19

Z

zymologist 19

Proper Nouns and Terms
（专有名词与术语）

A

accerlerating force　加速力
acrylic　丙烯酸纤维
African　非洲人
Alsace　（地）阿尔萨斯
aluminium　铝
amalgam　汞合金，汞齐（尤用于补牙）
Americas　美洲
Anatolia　（地）安纳托利亚
Andronovo sites　安德罗诺沃遗址
Annonay　（地）阿诺奈
antimony sulphide　硫化锑
antimony　锑
Anyang　（城）安阳
arsenic　砷
Ashkelon　阿什克伦（古城市名）
Assyria　（古国）亚述
Atlantic Ocean　大西洋
Aurignacian period　（考古）奥里尼雅克期（指法国旧石器时代前期）

B

Babylon　巴比伦；奢华淫靡的大都市
Babylonian　巴比伦人
Baker Street　贝克街
Baltimore　（城）巴尔的摩
Baluchistan　俾路支（巴基斯坦的一个省）
Battle of Ain Jalut　艾因贾鲁之战
Bavaria　巴伐利亚
bismuth　铋
Bolton　（城）博尔顿（位于英格兰西北部）
Borrowdale　（地）博罗代尔
Borsippa (Birs Nimrud)　波尔西帕（古代美索不达米亚南部苏美尔的重要城市）
Boston　（城）波士顿
British　英国人
Broadway　百老汇（美国纽约市戏院集中的一条大街）
Bronze Age　青铜器时代
Byzantine Empire　拜占庭帝国；东罗马帝国

C

carbolic acid　石炭酸，苯酚
carbon intermediate steel　中碳钢
cast iron　铸铁
Caucasus　高加索山脉
celerifere　塞莱里费尔自行车
cellulose pulp　纤维素纸浆
Central Asia　中亚
Central Europe　中欧
Chicago　（城）芝加哥
Chicago River　芝加哥河
chloroform　氯仿、三氯甲烷（旧时医用麻醉剂）
circular/twisting accerleration　圆周加速度
Coalbrookdale　（英）煤溪谷
coal-tar　（制作煤气时产生的）煤焦油，煤沥青
Colgate　（品牌名）高露洁
Colossus of Rhodes　罗得斯岛上的太阳神巨像
Cornwall　（英）康尔沃郡
course protractor　航向分角器
crank mechanism　曲柄机构
Croatia　（国）克罗地亚
cyanoacrylate　氰基丙烯酸盐
Cyprus　（国）塞浦路斯
Cyrus the Great　居鲁士大帝（波斯国王）

D

Derbyshire　（英）德贝郡
Deutschland　（国）德国；德意志号
draisienne　德莱斯式自行车

Dublin （城）都柏林
Dumfriesshire （英）邓弗里斯郡
Dupont de Nemours 杜邦公司

E

East Africa 东非
Eddystone Lighthouse （英国）埃迪斯通灯塔
Egyptian Blue 埃及蓝（硅酸铜钙）
Egyptian 埃及人
electro-chemical battery 电化学电池
Emperor Cheng 汉成帝
Emperor Huangdi 黄帝
English Law 英国法
Esperanto 世界语（1887年创制的一种人造国际语言，以欧洲主要语言为基础）
ether 乙醚
Ethiopia （国）埃塞尔比亚（旧称Abyssinia）
Etruria 伊顿鲁里亚（意大利中部的古国）
Etruscan 伊特鲁里亚人
Euphrates 幼发拉底河

F

falcata 短剑；利刃弯刀
Fangyan 《方言》
Feishayan (Flying Sand Weir) 飞沙堰

G

Gampi 雁皮树
Geneva （城）日内瓦
Genoa （城）热那亚（意大利港市）
Georgia 格鲁吉亚
Germany （国）德国
Gesher Benot Ya'aqov site (*abbr.* GBY) 格瑟尔·贝诺·雅科夫遗址
Great Bath 大浴池
Great Britain 大不列颠（包括英格兰、苏格兰和威尔士）
Great Pyramids 大金字塔
Greek 希腊人
Gregorian calendar 阳历；公历
Griffon Vulture 秃头鹫
gypsum 石膏

H

Hagia Sophia 圣索菲亚大教堂
Han Dynasty 汉朝
Han Empire/Dynasty 大汉王朝（汉朝）
Hanging Gardens 空中花园
Hermès （品牌名）爱马仕
Hohle Fels Cave 霍赫勒菲尔斯洞（位于德国，洞中文物是史前艺术和乐器的最早例子）
Holland （国）荷兰
Homo erectus 直立人（能用腿行走的早期智人）
Homo heidelbergensis 海德堡人（一种已灭绝的古老人类的物种或亚种，属于智人）
Hougang (site) 后冈遗址
Hudson River 哈德逊河
Hun 匈奴；匈奴人
hydraulic turbine 水轮机

I

Iberia （地）伊比利亚
Iberian Peninsula 伊比利亚半岛
Indian 印第安人
Indonesia （国）印度尼西亚
Indus 印度河
Industrial Revolution 工业革命
Inuit 因纽特人［单数是Inuk］
Iowa （美）爱荷华州
Iran （国）伊朗
Iraq （国）伊拉克
Iron Age 铁器时代
iron oxide 氧化铁
Isle of Crete 克里特岛
Israel （国）以色列
Italy （国）意大利

J

Japan （国）日本
Jiahu 贾湖（黄河流域新石器时代遗址）
Julian calendar 儒略历
Jupiter 木星

Proper Nouns and Terms
（专有名词与术语）

K

Kashmir　（地）克什米尔
King Wu Ding　武丁王

L

Lake Taihu　太湖
Lancashire　（英）兰开夏郡
Leather Lane　皮革巷
Leshan Giant Buddha　乐山大佛
Levant　泰凡特（地中海东部沿岸诸国和岛屿）
Liangzhu (site)　良渚遗址
lime　生石灰
limestone　石灰岩
linear accerleration　线性加速度
Liverpool　（城）利物浦
Louis Vuitton　（品牌名）路易·威登
Lyon　（城）里昂

M

Magdalenian period　（考古）马格德林期（欧洲旧石器时代的最后期）
Maillard reaction　美拉德反应
Majiabang culture　马家浜文化
Mamluks　马穆鲁克人
Manchester　（城）曼彻斯特
manganese oxide　氧化锰
Maqiao (site)　马桥遗址
Maryland　（美）马里兰州
Massachusetts General Hospital　麻省总医院
Mehrgarh　梅赫尔格尔（巴基斯坦的考古遗址）
Mesolithic Age　中石器时代
Mesopotamia　美索不达米亚
Mesopotamian　美索不达米亚人
Metal Age　金属时代
Mexico　（国）墨西哥
Middle Ages　中世纪时代
Milan　（城）米兰
Miletus　米利都（古代爱奥尼亚的城市）
Milky Way　银河
Ming Dynasty　明朝

Minoan Palace of Knossos　米诺斯王宫
Middle East　（地）中东
Minoans　（公元前2800—1100年前后以克里特岛为中心发达起来的）米诺斯文化的人
MIT　麻省理工学院（全称 Massachusetts Institute of Technology）
Mo Di　墨翟（即墨子）
Mohenjo-Daro　（古城）摩亨佐-达罗
Mongol　蒙古人
Morse code　摩斯密码

N

naphtha　石脑油（作燃料或用于制造化学品）
Naples　（城）那不勒斯
Neanderthal　（石器时代生活于欧洲的）尼安德特人
Near East　（地）近东（地中海东部沿海地区）
Neolithic Age　新石器时代
Netherlands　（国）荷兰
New York　（城）纽约
Niagara River/Falls　尼亚加拉河/瀑布
Nile　尼罗河
Nineveh　（古城）尼尼微
nitrous oxide　氧化亚氮，笑气（旧时牙医用作麻醉剂）
Noric steel　诺里克钢
North America　北美洲
Northern Cape　（南非）北开普省
Northern Wei Dynasty　北魏

O

obsidian　黑曜石；黑曜岩
ochre　赭石；赭土；赭色；土黄色
Olympia　（城）奥林匹亚

P

Pakistan　（国）巴基斯坦
Palaeolithic Age　旧石器时代
Palestinian Judaism　巴勒斯坦犹太教
Palmipède　巴拉米贝德号
Paris Exposition　巴黎博览会
Pendleton　（地）彭德尔顿
Penny Black　黑便士（英国1840年首次发行的便士

邮票，是世界上第一枚可粘贴在信封上的邮票）

Penny Post　（史）一便士邮政制
Pergamum　（古国）帕加马
Persia　（古国）波斯
Persian Empire　波斯帝国
Persian　波斯人
Phoenician　腓尼基人
phosphorus　磷
photovoltaic effect　光伏效应
pig iron　生铁；铸铁
Pisa　（城）比萨
plaster of Paris　熟石膏
polyamide　聚酰胺
polyester　聚酯纤维；涤纶
Portland　（城）波特兰
potassium chlorate　氯酸钾
primary cell　原电池，一次性电池
Pseudodoxia Epidemica　《流行的假知识》
puerperal fever　产褥热
Pyroscaphe　舰船号

Q

Qingli　庆历（北宋宋仁宗使用的年号）
quench-hardened steel　淬硬钢

R

Renaissance　文艺复兴（欧洲14、15和16世纪时，人们以古希腊罗马的思想文化来繁荣文学艺术）
River Saône　索恩河
River Seine　塞纳河
River Severn　塞汶河
rocker arm　（发动机，尤指内燃机的）摇臂
Roman Empire　古罗马帝国
Roman　罗马人
Romania　（国）罗马尼亚
Russia　（国）俄罗斯

S

saltpetre (potassium nitrate)　硝酸钾
Scandinavia　（地）斯堪的纳维亚
Schweppes　（饮用水品牌名）怡泉
Seljuk Turk　塞尔柱土耳其人

shaman　萨满
Shang Dynasty　商王朝
Shrewsbury　（英）什鲁斯伯里
Sibudu Cave　诗巫渡洞（位于南非，是中石器时代遗址）
Sicilian　西西里岛人
silicon nitride　氮化硅
Silk Route　丝绸之路
silver nitrate　硝酸银
Sintashta-Petrovka　辛塔什塔-彼得罗夫卡（文化）
solid electrode　固体电极
Song Dynasty　宋朝（分为北宋和南宋）
spandex　氨纶（弹性纤维）
Spartan　斯巴达人
spongy mass　海绵体
static charge　静电荷；静态荷
Statue of Zeus　宙斯神像
stibnite　辉锑矿
Stonehenge　（英国）巨石阵
sulphide　硫化物
sulphur　硫磺
Sumer　苏美尔（古地区名）
Sumerian　苏美尔人
Super Glue　（品牌名）强力胶
Swabian Alps　斯瓦比亚阿尔卑斯山脉
Syracuse　（城）锡拉库扎（位于意大利西西里岛东海岸）

T

Tanner Street　坦纳街
Teagle　提升机
Tepe Gawra　高拉土丘（古代美索不达米亚定居点，位于伊拉克北部）
the Midwest　（美）中西部地区
the Northwest　（美）西北部地区
the Queen　英国女王
the state of New York　（美）纽约州
the Stockton & Darlington Railroad Company　斯托克顿和达林顿铁路公司
Thebes　（城）底比斯

Proper Nouns and Terms （专有名词与术语）

Thirty Years' War　三十年战争
three-arm protractor/station pointer　三臂分度器
Thwaites' Soda Water　（饮用水品牌名）思韦茨苏打水
Tigris　底格里斯河
tin　锡
torquetum　赤基黄道仪
tropical year　（天）太阳年
tungsten　钨
Tuscany　（地）托斯卡纳

U

Ulm　（城）乌尔姆（位于德国南部）
Upper Canada　（地）上加拿大（指五大湖区以北和渥太华河以西主要讲英语的地区）
Upper Palaeolithic era　旧石器时代晚期
Ur　乌尔（古代美索不达米亚南部苏美尔的重要城市）
USS Nautilus　鹦鹉螺号

V

velocipede　早期脚蹬车

Venice　（城）威尼斯（意大利港市）
Vienna　（城）维也纳
Viking　维京人；北欧海盗

W

Warring States period　战国时期
Washington　（美）华盛顿州
Washington DC　（美）华盛顿特区
Welland Canal　韦士兰运河
Western Concert　（现代管弦乐）长笛
white lead　铅白（主要成分为碱式碳酸铅）
World Health Assembly　世界卫生大会
WPM　（单位）每分钟字数
wrought iron　熟铁

Y

Yixing County　宜兴县

Z

Zhibo Qu　智伯渠
Zhou Dynasty　周朝
ziggurat　塔庙（古代美索不达米亚的阶梯式金字塔形建筑）